# OSSEOINTEGRATION OF ORTHOPAEDIC IMPLANTS

# OSSEOINTEGRATION OF ORTHOPAEDIC IMPLANTS

AMIRHOSSEIN GOHARIAN

ELSEVIER

ACADEMIC PRESS

An imprint of Elsevier

Academic Press is an imprint of Elsevier
125 London Wall, London EC2Y 5AS, United Kingdom
525 B Street, Suite 1650, San Diego, CA 92101, United States
50 Hampshire Street, 5th Floor, Cambridge, MA 02139, United States
The Boulevard, Langford Lane, Kidlington, Oxford OX5 1GB, United Kingdom



**Library of Congress Cataloging-in-Publication Data**
A catalog record for this book is available from the Library of Congress

**British Library Cataloguing-in-Publication Data**
A catalogue record for this book is available from the British Library

ISBN 978-0-12-813384-2

For information on all Academic Press publications
visit our website at https://www.elsevier.com/books-and-journals



*Publisher:* Mara Conner
*Acquisition Editor:* Fiona Geraghty
*Editorial Project Manager:* Gabriela D. Capille
*Production Project Manager:* Anitha Sivaraj
*Cover Designer:* Greg Harris

Typeset by SPi Global, India

Cover credit: © Fraunhofer-Institut Fur Fertigungstechnik und Angewandte Materialforschung IFAM/Business Segment Medical Technology.

# DEDICATION AND ACKNOWLEDGMENTS


Dedicated to the spirit of *my father*

for his efforts to give love, aid, and cure to who were needy,

and special thanks to him

for teaching me to be kind and supportive to others, to be diligent and creative in achieving what I wish, and to have a strong spirit, high patience, and big hopes.

Dedicated to my merciful and lovely *mother* and special thanks

for her love and selflessness in my whole life,
for her patience and encouragement,
for teaching me to be righteous, moral, hopeful, and self-confident.

Dedicated to my lovely and elegant *wife* and special thanks

for her love and encouragement,
for her patience during writing of this book,
for her great, soothing, and merciful companionship.

Dedicated to my cute and elegant *daughter*

for her lovely presence in my world and making my life full of hopes, passions, and energy.

Special thanks to my supervisor *Prof. Ruslan Abdullah*

for his encouragement and help

Special thanks to *Elsevier publisher*

for effective and professional editing and publishing of this book

Special thanks to my supportive and sympathetic *brother*

for the good hopes and wishes.

Special thanks to my sympathetic *sisters*

for their good hopes and wishes.

Special thanks to my *relatives, friends, and colleagues*

for all their help and best wishes.

# CONTENTS

# ABOUT THE AUTHOR

**Mr. Amirhossein Goharian** is an independent product developer in the field of orthopedic implants. He earned a master's degree in both biomechanical and biomaterial engineering. With background in mechanics, biomaterials, and biomechanics, he has been engaged in research and development in the areas of trauma implants, spinal implants, and total joint replacements since 2011. His main concern and interest in the development of orthopedic implants is the effective incorporation of biomechanical, material, biological, and clinical facets simultaneously to optimize the implant design, technology, and features and ultimately to enhance the biomechanical and clinical benefits of orthopedic implants in the treatment of bone injuries, particularly in patients with osteoporotic bones. He attempts to challenge the current concepts and methods of evaluation and development of orthopedic implants and to create and propose novel concepts for future development of these products. After 5 years of R&D engineering in the industry of Trauma Plating Systems and publishing of the book "Trauma Plating Systems" with honored and greatly professional publisher "Elsevier," he has initiated product development in the industry of Spinal Implants and Total Joint replacements. One of the key topics in R&D engineering of these types of implants is the osseointegration aspects of the implants in long-term implantation. This has motivated investigations in the field of osseointegration for use in bioactivity enhancement of the orthopedic implants and publishing of "Osseointegration of Orthopedic Implants" as the second author's book in the area of orthopedic implants. The author would establish the novel conceptual plating system as "AHealFS" in previous book and explore this system for presenting the "Transition Layer" concept for effective mechanical and biological integration of the bone to the implant in longer period of implantation compared to the current osseoconductive conception methods. From his viewpoint, the combination of the R&D engineering and authoring allow for significant enhancement in scientific product development according to the innovative achievements and techniques and substantial enhancement in authoring of scientific-engineering books in the area of orthopedic implant industry.

# PREFACE

Historically, integration of the implant to the bone tissue has been a major challenge for effective internal implantation. Due to the compact and high mechanical properties of the cortical bone, gripping of the implant to the cortical segment of the bone would be achievable. In fact, cortical bone acts as the tough material to secure fixation of the implant. However, tapping of the implant to the cortical bone might be affected by physiological dynamic loading conditions which would be created through normal daily activities. In some cases, the thickness of the cortical bone is not sufficiently large to provide effective tapping or gripping of the implant for long-term resistance to dynamic shear stress between the bone and the implant. Furthermore, tapping strength of the implant to the low mineral density bone would be significantly reduced compared to the natural compact cortical bone. The risk of implant migration, loosening, or dislocation is increased in cases with poor or ineffective tapping of the implant to the cortical bone. In majority of the implantations, the implant is purchased through both the cortical and the trabecular bone. Purchasing through the cortical bone would provide fixation strength while purchasing through the trabecular bone would enhance stability of the implantation. Integration of the implant with trabecular bone would be crucial to enhance the shear strength at the interface of bone-implant. The trabecular bone is highly affected after the implantation of an artificial device and the rate of the bone resorption at the peri-implant zone is significant. The response of cortical bone at the peri-implant zone is thickening of the cortical section of the bone tissue. Therefore, enhancement of osseointegration of the implant at the trabecular bone is greatly crucial for effective and long-term implantation of the orthopedic implants. Nowadays, various types of processing methods are utilized to enhance the bioactivity of the implant surfaces which are in direct contact with the bone, particularly trabecular bone. In majority of these methods, an osseoconductive layer or structure is established to conduct the adherence of osteogenic cells and the growth factors, thus its proliferation and differentiation on the implant surface. This conductivity would be enhanced with flow of blood through the osseoconductive layer for mineralization and maturation of the created osteocyte cells to be formed as woven bone tissue and ultimately as lamellar bone. The concepts of osseointegration in various types of orthopedic implants might be different in some specific aspects.

According to the type of implantation and treatment of bone injuries, integration of the bone and implant would have specific mechanical and biological considerations. For instance, in placement of cementless prosthesis in total replacements, thickness of osseoconductive layer in microscale on the implant surface is desired, while in placement of titanium (Ti)-coated fusion cages in spinal fusion surgery, thickness in the nanoscale with higher adhesion strength is advantageous. Thus, such biological and mechanical requirements would necessitate various types of processing methods for the creation of osseoconductive or osseointegrated layer on the surface of the implant.

With consideration of various aspects of osseointegration in orthopedic implants, having a compact book in this topic has been found necessary to be written to review, discuss, and challenge the current concepts of osseointegration in the development of orthopedic implants which is now available as *"Osseointegration of Orthopedic Implants"*. In first section, the fundamental of osseointegration along with implant loosening are expressed to clarify the main concepts of osseointegration of orthopedic implants. Likewise, the effective mechanical and biological parameters that would influence the promotion or deterioration of the osseointegration are explained. In the second section, the orthopedic implants (categorized as trauma fracture fixation implants, total joint replacements, spinal implants, and bone tissue scaffolds) are presented in general characteristics along with specific aspects of osseointegration in each category of the implants. The concepts of osseoconductivity are reviewed in third section to explore the development challenges for enhancement of osseointegration. Based on the explored aspects of osseoconductivity, the main industrial processing techniques that are utilized commercially or has potential advantages to be marketed, are reviewed and discussed afterwards. Coating or layering of bioactive or osseoconductive materials on the surface of the orthopedic implants are currently two main techniques for enhancement of the osseointegration. Expression of these techniques along with their challenging aspects and long-term effectiveness are provided in this section. In the final section, conception of the osseointegration in the development process of the orthopedic implants is represented. In this section, the involved mechanical, biological, and clinical aspects of the bone-implant integration are deeply elaborated to provide comprehensive outlook of bioactivity enhancement of the implant in contact surfaces with the bone. Furthermore, engineering aspects of osseointegration in conjunction with evaluation methods are discussed to explore the effects of processing and testing methods on effectiveness of the osseointegration in real physiological conditions. In the last chapter of the book, the concept of transition layer as the new technique of osseointegration enhancement is presented and discussed from which some novel ideas (e.g., bulk-porous transition layer, transition-osseoconductive transition layer, and Mg-PEEK transition layer) have been hypothesized.

Overall, *"Osseointegration of Orthopedic Implants"* would be a compact reference for various aspects of the osseointegration and associated development strategies, techniques, and methods for mechanical and biological enhancement of the orthopedic implants in contact surfaces with the bone. This book is recommended to biomedical researchers, mechanobiological investigators, implant developers, and particularly orthopedic and neurospine surgeons to get familiar with osseointegration benefits of the orthopedic implants in the enhancement of clinical outcomes. All relevant audiences and readers are welcome to share their notes and comments with the author via *orthoimplant.develop@gmail.com* for improvement of the book in the next editions. The author would like to offer technical cooperation in the topic of the book with the institutes which are interested in enhancing the bioactivity characteristics of the orthopedic implants in contact surfaces with the bone.

## CHAPTER 1

# Fundamentals in Loosening and Osseointegration of Orthopedic Implants

## 1.1 INTRODUCTION

The concept of osseointegration was first developed for dental implants to address the in-growth and on-growth of bone tissue to the surface of titanium dental implants. Gradually, this biological advantage was addressed for the development of orthopedic implants and for the evaluation of implant bioactivity aspects. In majority of implantation cases, osseointegration of implant is greatly beneficial; however, there would be some cases at which poor osseointegration is desired. Generally, better osseointegration of implant could enhance implantation of the implant to the bone, which results in better treatment of bone injuries. It has been clinically reported that poor integration of bone and implant would cause loosening of implant or even suppression of the bone to the implant.

Osseointegration is a scientific term that is going to be of a practical-scientific development concept [1–9]. The understanding of interaction between the bone and the implant has opened various fields of researches and investigations. Undesired biological response of human body after the implantation of implant was the initial side effect of internal treatment utilizing various types of materials in fabrication of implants. Therefore, the standard "ISO 10993" was developed to determine the essential biological requirements for the biological evaluation of materials aimed to be used in fabrication of implants. It would be very crucial to assure no biological effects (e.g., toxicity, cytotoxicity, sensitization, and genotoxicity) of implant materials in human body fluid and in contact with bone tissue for the development of orthopedic implants. In fact, implant must be without chemical reaction in ionized body fluid or blood if undesired release of implant materials causes biological signs or symptoms. Other factors such as low bacterial adhesion have also been highlighted to enhance biological safety or biocompatibility of the implant.

*Osseointegration of Orthopaedic Implants*
https://doi.org/10.1016/B978-0-12-813384-2.00001-1

In majority of the implantations, the bonding of the implant to the bone is strongly beneficial. Indeed, good biocompatible implant without enhanced clinical benefits would not be a superior solution for the treatment of bone injuries. It has been observed that without effective bonding or integration of orthopedic implants to the bone tissue, various types of clinical complications arise postoperatively. Thus, bone-implant integration with high level of shear strength under dynamic loading conditions (which is transferred to the bone-implant interfaces) is the key factor of implant stability inside the bone and ultimately enhances clinical outcomes. Ineffective osseointegration of implant and bone may result in loosening or dislocation of the implant from the original implanted location [10–20]. Poor integration of bone and implant may occur because of various reasons. In fact, whatever would affect the bone resorption and deposition process at the bone-implant interface could cause loosening of integrity between the bone and the implant.

In this chapter, first the history of osseointegration concept is reviewed to clarify how this biological interaction could enhance the clinical results. Then various parameters that could influence the better integration of bone-implant are elaborated. Following that, the differentiation and formation of bone tissue at the bone-implant interface is discussed, then the effective expression of osteoblastic genes in promoting osseointegration is expressed. In second part of this chapter, the loosening of the implant is reviewed from various aspects. The effects of contact wear, the level of mechanical stimulation, and implant-related factors on implant loosening are discussed. Due to high risk of implant loosening in total joint replacements (TJRs), the interaction between the bone cement and the bone is deeply expressed and development concepts that could reduce the risk of implant loosening are highlighted in general.

## 1.2 HISTORY OF OSSEOINTEGRATION

Osseointegration was first introduced by Professor Per-Ingvar Branemark in the 1950s as the attachment of human bone cells to the surface of metallic implants. Initially, reaction of the bone to the titanium alloy was seen in investigations carried out by Beaton and Davenport in 1940 and Gottlieb Leventhal in 1951. Through their research, potential use of titanium in orthopedic and dental implants has been addressed due to good strength, no body reaction, and growth of bone tissue on the surface of the titanium. In 1952, Per-Ingvar Branemark implanted a titanium chamber to the rabbit bone to investigate the effect of blood flow on progression of bone formation on the surface of the titanium chamber. After completion of the test, when he wanted to remove the chamber, he recognized that the chamber had integrated with bone and could not be removed easily. At that point, he discovered bone growth into the titanium chamber and good integration of bone and titanium implant. After that, he called this phenomenon "osseointegration" to be used as a material characteristic to be used in the development of implants for the better treatment of bone injuries. Later on in 1965, Branemark developed dental implant using the osseointegration ability of the titanium. Based on the description of Nishimura [5] from Branemark's definition, osseointegration is stable connecting of the bone and implant without sign of infection, inflammation, and clinical undesirable issues. Today, the concept of osseointegration is used extensively in the development of orthopedic, spine, and dental implants to promote bone in-growth and on-growth through the implant. Osseointegration may start with bone resorption at the interface of bone-implant, then formation of bone tissue on the implant surface or through in the implant body, and ultimately integration of the bone and implant at the contact surfaces.

## 1.3 EFFECTIVE PARAMETERS ON IMPLANT OSSEOINTEGRATION

Integration of orthopedic implants and bone tissue is a very complex phenomenon. Bone tissue is inherently a smart bioactive structure at which alteration of its architecture, density, size, and composition are permanently changing based on the mechanical, chemical, physical, hormonal, and stimulation signals for proliferation and differentiation. On the other hand, orthopedic implants are normally made of bioinert materials with no bioactivity characteristics. Implantation of implants with bioinert material to the bone would cause alteration of bone remodeling process at the bone-implant interface and if this abnormal bioactivity of bone tissue is continued, more depth of bone tissue at the bone-implant contact area is affected. In fact, this abnormality may increase resorption of bone tissue by osteoclast, which results in weak or no integration of bone through the implant surface. Ineffective integration of implant to the bone would affect clinical outcomes significantly. Therefore currently the topic of osseointegration is the main development concept to enhance the effectiveness of the orthopedic implants. This topic has been addressed as crucial concern to reduce the clinical complications, however, implant with high osseointegration characteristics (osseointegrant) for internal implantation treatment is not commercially available.

In terms of orthopedic implants, manufacturers attempt to enhance the surface conditions of available commercial implants (made of bioinert materials, e.g., titanium alloy, polyetheretherketone (PEEK), stainless steel, cobalt-chromium alloy) for better penetration and formation of bone tissue on the implant surface. This development concept is actually the only concept that has been addressed for the enhancement of implant bioactivity. Some materials are inherently allowing penetration and formation of bone tissue on their surface (e.g., Ti alloy) while some are naturally weak in integration with bone tissue. Several surface parameters influence osseointegration ability of the implants. Surface roughness, bacterial adhesion, passivation layer, surface oxidation, corrosion resistance, porous nature of surface, etc., are some of these parameters. In addition, mechanical stimulation signals at the interface of bone-implant (will be reviewed in Section 1.8) would be another factor that affect or enhance the formation of bone tissue and its maturation after implantation. Furthermore, the morphology and density of the bone tissue at the implanted zone is effective on induction of bone tissue to be integrated with implant. In this case, the design of implant needs to be compatible with the bone mass and architecture. The other effective factor in osseointegration is the local circulation of human body fluid at the implanted area and nontoxic or chemically affected blood at the bone-implant interface. Undesired conditions of human body fluid may have been caused due to the implant material release, infection, or other biological effects that reduce the effective mineralization and proliferation of bone cells at the bone-implant interface.

## 1.4 PRIMARY AND SECONDARY STEPS OF OSSEOINTEGRATION

Integration of an implant to the bone could be evaluated in two steps as primary and secondary integration. The primary integration is the mechanical interlocking of the implant inside the compact cortical and high porous trabecular bone and the secondary integration is biological bonding between the bone and the implant. It means that for an effective osseointegration of implant, these two steps of integration are desired. It could be said that in the implantation of implant through the high porous bone without engagement with cortical bone (e.g., implantation of knee formal, knee tibial tray, hip femoral stem, etc.), the primary stability of the implant for secondary biological integration may not be achieved [in the implantation of total joint implants, normally the implant is inserted through the trabecular bone (e.g., knee femoral and knee tibial tray) or is in contact with cortical bone (hip femoral stem)]. In fact, primary mechanical integration enhances the stability of the implant to allow biological bonding at the implant-bone integration. Although, such implants are not interlocked with the compact cortical bone, they would be constrained by surrounding cortical bone and stabilized by bone cement from which the implants would not have freedom in the axis of loading conditions and this could be considered as the primary mechanical constraints.

## 1.5 ENDOSSEOUS CONCEPT IN PROMOTING OSSEOINTEGRATION

Based on Branemark's observations [21], in-growth of bone tissue into the implant having internal holes or dent was reported. Recently, endosseous implant has been developed in the treatment of bone injuries such as endosseous screw in fracture fixation of foot metatarsal bone. The concept of endosseous implant was also utilized for the development of dental implants. In a recently developed osseointegrated implants (e.g., leg and finger osseointegrated implants for implantation through the femur and metacarpal bones, respectively, in patients who have lost the leg and the finger), the stability of the implant is established by in-growth of compact bone tissue through the internal body of the implant. It is advantageous to allow the growth of the bone tissue inside the implant body through the created holes or empty spaces. By compacting of the bone graft materials (autograft, allograft, xenograft, or synthesized graft) or structures (bone tissue scaffold, bone substitution graft) through these internal holes, the biological bridge between the bone tissue surrounding the implant could be achieved by which the bone tissue can penetrate through the grafting structure and create an integrated bone structure with the surrounding bone at the implanted zone. The bone-implant internal integrating would provide long-term stability of the osseointegrated implants with low level of dislocation or loosening complications as that might be observed in nonosseointegrated implants (majority of the current orthopedic implants such as TJRs, trauma plating systems, spinal screw-rod stabilization are nonosseointegrated implants which has external integration with the bone tissue through their surfaces). The concept of endosseous implants have been considered for the development of dental implants and spinal intervertebral or vertebral cages for many years, however, this concept is currently under development for the treatment of severe bone loss injuries or in patients with an amputation in various parts of the body (mostly in hand and leg organs) as the new method for the replacement of the partial or whole body organ such as leg, hand, finger, or foot (one type of leg osseointegrated implant is shown in Fig. 1.1).

## 1.6 EXPRESSION OF NONCOLLAGENOUS MOLECULES OR GENES

Gene expression is a biological term to express the conversion of DNA to a functional signal to react against of changes in surrounding environment of the cell. This functional signal is formed as protein or molecules to induce or initiate a biological action. Noncollagenous molecules would facilitate differentiation of osteoblast for the formation of bone extracellular matrix (ECM). Therefore, the recapitulation of these molecules (osteocalcin, osteonectin, osteopontin, and bone sialoprotein) at the interface of bone-implant is the key factor for effective secondary osseointegration and their expression would increase differentiation, proliferation, and adhesion of the osteoblast through the surface of implants from which the level of matured bone at the bone-implant interface is increased. Nishimura [5] highlighted that if the state of gene expressing [22–26] of the molecules is stable or steady, osteoblast differentiation is effectively achieved at the interface of bone-implant [6, 27–30]. It has been also pointed that early expression of inflammatory cytokine-associated genes would enhance the formation of bone at the implanted site at the bone-implant interface [5, 31]. Osteoclast or cytokine genes would allow expression of these cells for resorption of the cut or affected bone tissue at the bone-implant interface to allow generation or creation of new bone tissue at the resorbed zone. Manjo et al. [26] reported that if the state of mRNA genes after implantation were steady, the integration of bone tissue to the implant would be improved. The sequential expression of the effective genes on formation of bone tissue at the interface of bone-implant has been well addressed by other researchers [32–34]. Other than osteoblastic genes, the expression of cartilage ECM genes has been observed after the implantation of the implant in the peri-implant bone tissue which enhances the mechanical property of the formed bone.

Osteoblasts as osteogenesis cells include the bone formation genes. It means that if osteoblastic genes are induced, generation of molecules or

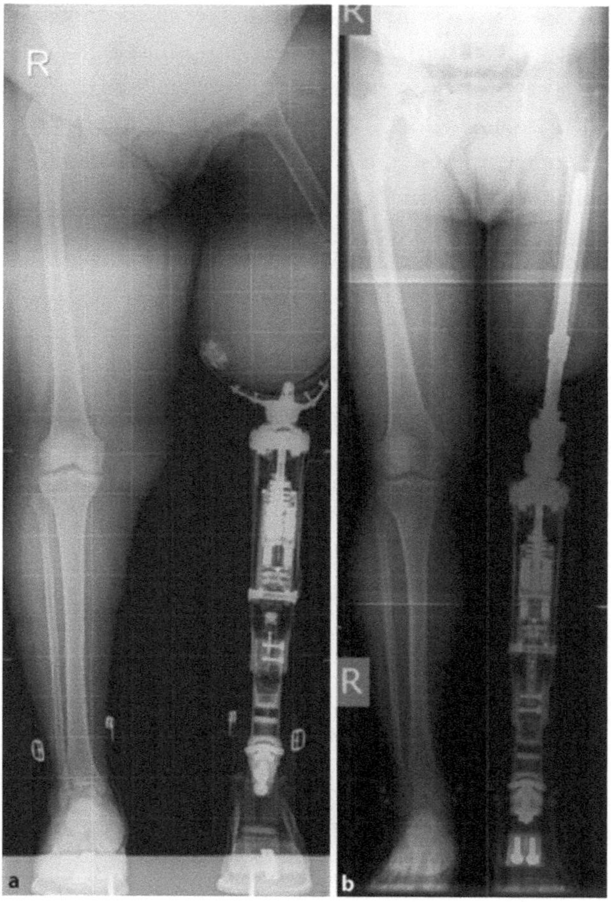


**Fig. 1.1** Implantation of the (A) socket implant, (B) osseointegrated implant, through the femur bone for the treatment of a transtibial amputation. In the osseointegrated implant, the osseointegrated part of the implant (the implanted long arm to the femur bone) is inserted through the intra-medullary hole of the femur bone. This implanted section of the whole prosthesis could have intra-medullary hole or coated with macro-porous titanium or tantalum with thickness of about 1.5 mm. In the former case, the osseointegration period would be about 6–12 months compared to about 2–3 months for the latter case. *(Courtesy of J. Frölke, R. Leijendekkers, H. Van de Meent, Osseointegrated prosthesis for patients with an amputation, Unfallchirurg 120(4) (2017) 293–299.)*

proteins for converting osteoblast cell to osteocyte cell would establish formation of bone tissue. As discussed in Section 1.8, one of the main parameter to induce osteogenic cells for bone formation is the alteration of mechanical loading conditions. This mechanical alteration would influence expression of osteoblastic genes from which the osteogenic cells are induced to initiate the process of bone formation. The other factor is hormonal changes which may affect the expression of osteoblastic gene and reduce the density of osteoblast cells at the intended zone. When an implant is implanted through the bone tissue, contact interfaces of bone-implant are resulted. This would impose a significant change in the surrounding environment of the bone tissue at the bone-implant interfaces. This change would greatly induce the expression of osteoclast-related genes for resorption of dead or severely affected bone matrices. Then due to the mechanical changes at the peri-implant interfaces (peri-implant is referred to the parts of the bone which are cut or loaded by the implant), the osteoblastic genes are also induced to be expressed. Transferring the alteration of loading conditions to the osteoblast cells would be other biologic matter that would be enhanced by promoting localized mechanical stimulation factors and good formation of primary collagenous tissue on the surface of the implant. Accumulation of blood or flow rate of blood or body fluid at the implanted site would also effect the mineralization of osteoblast cells to be formed as bone ECM at the bone-implant interface. There would be other biological factors that are involved in bone remodeling process, and the highlighted factors in this section are some of the factors that are mainly measured for the evaluation of bone tissue formation conditions at the bone-implant interface.

## 1.7 EFFECT OF CONTACT WEAR PARTICLES ON IMPLANT LOOSENING

The generation of particles due to the contact wear in joint replacement prosthesis could cause loosening of the implant. The presence of these particles (which are polymeric-, metallic-, or ceramic-based particles) at the implanted site would induce osteoclast cells to be formed as mature osteoclast to counteract the effect of wear particles. Increase in mature osteoclasts at the bone-implant interface promotes bone resorption from which the balance between bone resorption and deposition is interfered. In addition, accumulation of the macrophage precursors resulted in the resorption of the released particles which increases the risk of apoptosis of osteoblast cells, as a result the formation of new lamellar bone is significantly reduced. After the release of particles from the surface of implant (particles could be detached from the implant surface due to corrosion, fretting corrosion, fatigue corrosion, contact wear, galvanic corrosion, etc.), particle phagocytosis is initiated by macrophages which increases differentiation of the osteoclast. In fact, through the phagocytosis process, the apoptosis of released particles of the implant occurs and the remnants of this process could cause inflammatory response and osteolysis at the bone-implant interface [15, 19, 35] (Fig. 1.2 demonstrates the loosening of total knee replacement due to osteolysis as a result of polyethylene wear particles). The biological solution of this issue is the presence of a type of adipocytokine [adiponectin (APN)] to update the remnants of the cells from the apoptotic process of particles. Indeed, APN receptors would provide this adicytokine at the affected area to prevent osteolysis effect of apoptosis of implant released particles [36, 37]. On the other hand, if the activity of apoptosis is decreased, it would influence bone resorption of old bone at the interface of bone-implant which prevents the formation of new bone tissue between the bone and the implant. Landgraeber et al. [11] performed in vivo cell culture along with animal and human examination to study the influence of APN on the

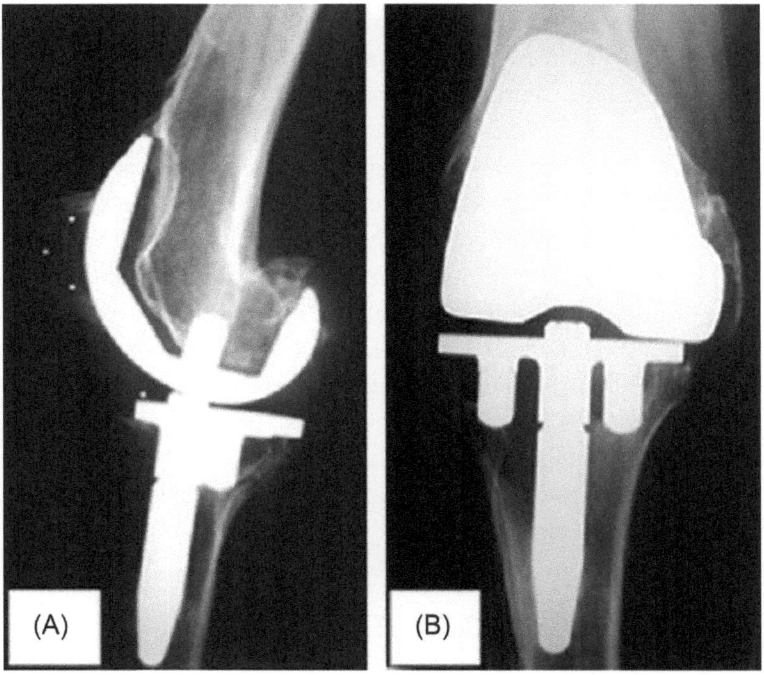


**Fig. 1.2** Loosening of femoral component in implantation of total knee replacement due to osteolysis as a result of polyethylene wear particles, (A) lateral view, (B) AP view. *(Courtesy of M. Bahraminasab, et al., Aseptic loosening of femoral components–a review of current and future trends in materials used, Mater. Des. 42 (2012) 459–470 and get permission from Elsevier.)*

prevention of implant released particles to induce osteolysis. They found significant expression of APN in the conditions aseptic loosening has occurred in animal and human cases. Likewise, significant expression of APN was detected in cell culture test in which released particles were introduced to the culture. As suggested by Landgraeber et al. [11], supplement of higher amount of APN than biologically supplied could enhance the removal of apoptosis remnants and inhibit osteolysis and loosening of the implant.

## 1.8 OVERSTIMULATION OF MECHANICAL LOADING EFFECT ON IMPLANT LOOSENING

Loosening of the implant might be due to overstimulation of loading conditions at the bone-implant interface. This type of failure has been reported for glenoid prosthesis in total shoulder arthroplasty (TSA) [13, 38–44] (Fig. 1.3). In shoulder joint (glenohumeral joint), the loading effect of rotator cuff and subscapularis muscles are significant on proximal portion of humerus bone, particularly humeral head. The glenoid component of TSA prosthesis is inserted into the proximal humerus and attached to the glenoid aspect of the scapula bone. The loading conditions on glenoid component are exceeded when the hand has flexion-extension and varus-valgus movement at the shoulder joint. The presence of such significant force would cause overinduction of mechanical signals which increases the expression of osteoclast-associated genes and resorption of the bone tissue at the bone-implant interface. In fact, after resorption of old bone at the micro-distance of implant, new bone tissues would not be formed and when the depth of bone resorption is increased to further distance from the implant, bone-implant gap is increased resulting in the loosening of the implant. Therefore a considerable number of revision surgeries of TSA have been performed [13]. It could be said that implantation of the implant or prosthesis to the bone must be managed in such a way that its stable integration during dynamic physiological conditions is achieved in long-term implantation.

## 1.9 FAILURE IN THE INTEGRITY OF BONE CEMENT AND BONE

Due to the presence of high range of loading conditions on TJRs, over displacement or strain of prosthesis components is inevitable. In order to increase the stability of TJR, bone cement [e.g., polymethyl methacrylate (PMMA)] is used to insert the prosthesis through the bone. The aim is

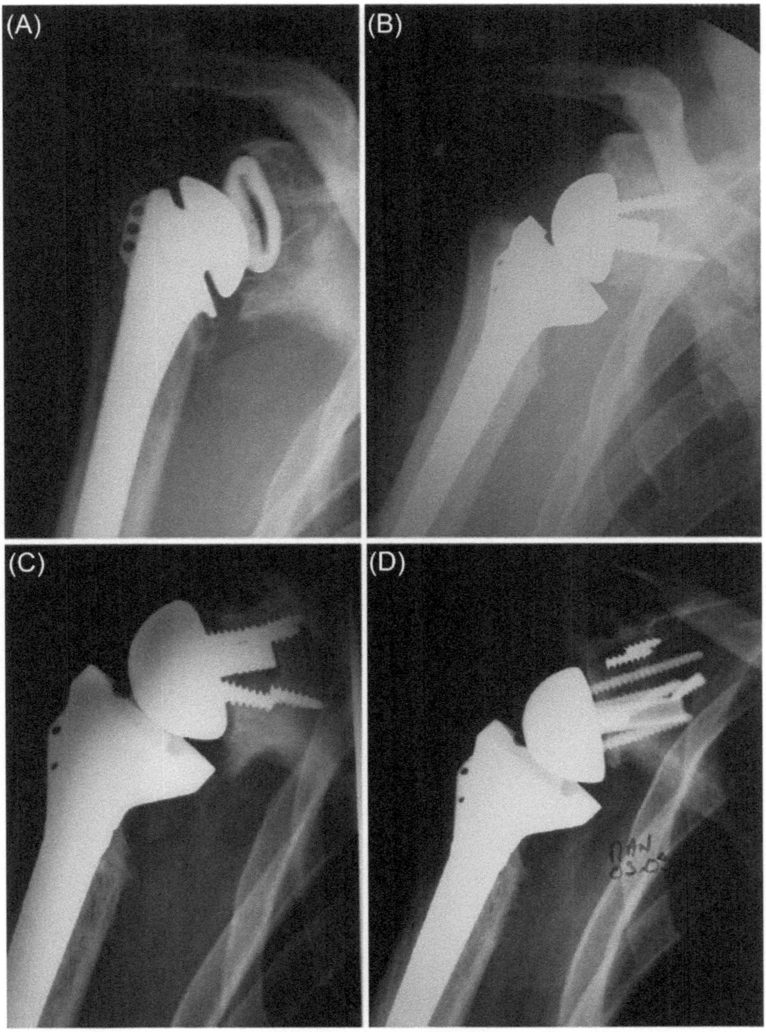


**Fig. 1.3** Implantation of total shoulder replacement (TSR): (A) loosening of the glenoid component due to overstimulation of mechanical loading, (B) revision by other type of TSR with glenoid component having a short peg including bone graft which is implanted through the glenoid portion of the scapula bone, (C) re-loosening of the revision TSR due to short length of the bone grafting peg, (D) reimplantation of the glenoid component with a long peg beyond the bone grafting spacing of the peg to increase the stability of the glenoid implantation under physiological loading conditions between the bone and the glenoid component. *(Courtesy of B. Melis, et al., Glenoid loosening and failure in anatomical total shoulder arthroplasty: is revision with a reverse shoulder arthroplasty a reliable option? J. Shoulder Elb. Surg. 21(3) (2012) 342–349 and get permission from Elsevier.)*

the integration of bone and cement to establish bone-implant integration. However, due to brittleness chemical behavior of the cement, the integrity of implant and cement might be affected in the long-term use [13]. Generally, deterioration of the shear strength at the interface of bone-cement could induce loosening or migration of the joint replacement prostheses [16] (Fig. 1.4). This failure occurs due to the dynamic physiological loading conditions at the treated joint. In fact, generated shear stress at the interface could detach newly formed bone tissue at the interface of bone-cement [18, 20, 45–48]. Interdigitation of the cement through the bone during joint replacement surgeries was highlighted to be the key factor for an increase in fatigue shear strength of bone-cement interface [18, 20, 47, 48]. To achieve optimal interdigitation, contact surface of the bone could be textured or roughened by using specific cutting tools. With an increase in bone-cement interdigitation, the contact stress would be increased which would promote the contact wear and generation of cement debris at the bone-cement interface. Therefore, the degree of surface roughening of the bone needs to be optimal to enhance fatigue shear strength while preventing the risk of wear debris generation at the interface.

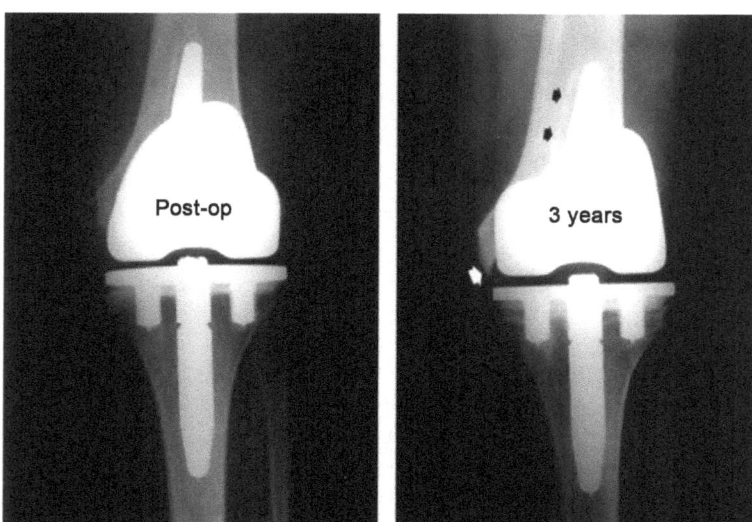


**Fig. 1.4** Migration of femoral component in implantation of total knee replacement after 3 years of the follow-up. *(Courtesy of M. Bahraminasab, et al., Aseptic loosening of femoral components–a review of current and future trends in materials used, Mater. Des. 42 (2012) 459–470 and get permission from Elsevier.)*

## 1.10 RELATION OF SYNOVIAL FLUID WITH IMPLANT LOOSENING

The relation of synovial fluid with loosening of implant is other biological factor that has been studied [17, 49–54]. It was discovered that synovial fluid acts as a carrier for induction factors of bone formation and resorption. However, based on the reported observation, synovial fluid has been detected around the loosened prosthesis which would be body reaction to the resulted inflammatory. The influence of overproduction of synovial fluid at the bone-implant interface increases the collagen expression factor (e.g., collagen I), resulting in brittle bone without effective mineralization and ultimately lamellar or mature bone tissue would not be formed.

## 1.11 LOOSENING OF TOTAL HIP REPLACEMENT

Loosening of total hip replacement has been extensively reported in the literature [12]. Due to remarkable number of revision surgeries due to aseptic loosening, researchers have performed various biological investigations to study the body reactions in cases with loosened THR prosthesis [55–57]. As highlighted in Section 1.7, the main cause of such implant loosening is body reaction to the foreign particles, which would not be biologically safe to be degraded without undesired outcomes [58–60]. Release of polyethylene (PE) particles is the major stimulation factor of macrophages differentiation to absorb these particles. It was highlighted that PE particles less than 5–7 μm are in the size range of macrophages and can be absorbed by macrophages [59, 61]. However, for larger particles size, the formation of giant cells is induced [62, 63] which cause creation of granuloma at interface of bone-implant [15, 60, 63–65].

## 1.12 IMPLANT-RELATED FACTORS FOR THE PROGRESSION OF IMPLANT LOOSENING

The geometry of the implant has been found effective in the early osseointegration of the implant [14, 66, 67]. It has been highlighted that several reasons cause bone loss during implantation of the implant such as drilling force, drilling rising temperature, or other technical overloading during surgery that can result in mechanical and thermal damage to the peri-implant bone tissue [68, 69]. Due to the creation of such bone losses, the magnitude of transferred strain to the bone tissue at the bone-implant interface may increase over than that of essential physiological strain for bone remodeling. As reported in the literature [10, 70], strain values in the range of 1000–1500 με would physiologically stimulate expression of related genes for bone remodeling, particularly the osteoblastic genes. This is while the strain in the range of 1500–3000 με would reduce the rate of bone remodeling and over 3000 με is associated with pathologic overloading and reduction in structural strength of bone tissue. The extent of bone losses at the interface of bone-implant would influence on magnitude of transferred strain to the bone. Based on the simulation and clinical observations [10] bone loss at the bone-implant interface would induce rise in strain which would affect osseointegration of the implant. Furthermore, early mechanical interlocking or integration of implant to the bone enhances biological integration at the bone-implant interface which would replace old bone tissues with new deposited bone tissue through the surface of implant. This scenario becomes significant at the interface of implant and cortical bone. Based on the achieved results by Wirth et al. [14, 67], load transferring from cortical bone through the trabecular bone has crucial impact on the creation of loading magnitude on trabecular bone. In other words, if the cortical bone-implant integration is not adequately achieved in early implantation, high magnitude of strain could be transferred to the trabecular bone which may cause weakening of the trabecular bone structure at the interface of bone-implant. This is the key point of cortical screw development at which fine threads with higher inner diameter was created at the screw neck in contact with cortical bone. The fine thread with small pitch would increase trapping of the screw into the cortical bone, which results in the enhancement of load bearing conditions in safe magnitude (1000–1500 με), and reduce transferring load value to the trabecular bone.

## 1.13 SUMMARY

Osseointegration of the orthopedic implant has been reviewed in general concepts in this chapter. In the first part, the main mechanical and biological factors that have effect on enhancement or deterioration of the osseointegration were reviewed according to the well-published recent studies. As highlighted in Section 1.2, the osseointegration concept as a character or factor was first introduced by Branemark for the development of dental implants for higher level of integrity to the mandibular bone. Due to the implantation of internal implants to the bone, the concept of osseointegration has been gradually utilized in the development of orthopedic implants to enhance the clinical outcomes in early and long-term implantation. In this regard, various types of mechanical and biological parameters could be considered in designing or processing orthopedic implants for the promotion of osseointegration as reviewed in Section 1.3 and summarized in Table 1.1. Primary and secondary concepts of osseointegration were then expressed. Initial mechanical

**Table 1.1** Effective parameters on osseointegration of the orthopedic implants

| Parameter | How to effect |
|---|---|
| Bone as a bioactive structure | Bioactive nature of the bone tissue as one of the body tissues would be a positive factor for the enhancement of implant osseointegration. Bone tissues could be restructured or reconstructed through the bone remodeling process in which various types of biological factors are involved to facilitate regeneration of the bone tissue and resorption of the dead or old bone tissue at the interface of the bone and the implant. Therefore, such bioactive character of the bone tissue would allow adoption of the bone tissue with the implant. |
| Implant as a bioinert structure | Implants fabricated from bioinert materials (nearly all of the current available orthopedic implants with a high implantation rate in the treatment of bone injuries are manufactured from bioinert metals, ceramics, or polymers) are totally inert in the human body and therefore could be a negative factor for the enhancement of osseointegration. Inertness of the implant would not allow its interaction with the bone tissue to enhance the adoption of the bone and the implant at the contact surface, because of which there would not be an effective mechanical interaction between the bone and the implant. Likewise, no biological interaction could be established between the bone and the implant at all, which would cause weakening or loosening of the press-fit bonding of the bone and the implant in long-term implantation. |
| Implant material | Inertness of the implant might be effected by the chemical reaction of the implant with the human body fluid, which would be a negative factor for the enhancement of osseointegration. Through this reaction, the undesired particles would be released |

*Continued*

**Table 1.1** Effective parameters on osseointegration of the orthopedic implants—cont'd

| Parameter | How to effect |
|---|---|
| | into the human body fluid and induce immune or macrophage cells to resorb the released particles from the surface of the implant. This phenomenon would interfere with the biological process of bone remodeling and increase the activity of the osteoclasts in the resorption of the bone at the bone-implant interface. |
| Implant surface conditions | The conditions of the implant surface could be a negative or a positive factor for the enhancement of osseointegration. This means that by processing of the implant surface through effective methods, the bone-implant integration could be increased. This is the parameter that is under control of the implant developers to reduce the mechanical and biological inertness of the implant in interaction with the bone. Currently, majority of the development concepts in the enhancement of osseointegration are investigated in this area. The surface characteristics such as surface roughness, bacterial adhesion, passivation layer, surface oxidation, corrosion resistance, surface porosity, porous pore size, bioactive coating layering, osseoconductive coating, layering, etc. are contemplated for promoting bone formation on the implant surface. |
| Mechanical stimulation signals | Mechanical stimulation signals at the bone-implant interface could be a negative or a positive factor for the enhancement of osseointegration. Severe mechanical loading would generate excessive mechanical signals from which induction of the osteogenic cells is significantly affected. Conversely, a low level of mechanical loading could not generate sufficient mechanical signals for the stimulation of osteogenic cells to be formed as a mature bone and to strengthen the bone extracellular matrix at the bone-implant interface. The mechanical mismatch between the metallic implant and the bone would increase the risk of stress shielding at the bone-implant interface from which majority of the loading is borne by the implant and mechanical stimulation signals would not be sufficiently generated. |

**Table 1.1** Effective parameters on osseointegration of the orthopedic implants—cont'd

| Parameter | How to effect |
|---|---|
| Bone morphology and density | Morphology and density of the bone tissue at the implanted area would be considered in the design of the implant. Therefore, it could be a positive factor in the enhancement of osseointegration. Indeed, the morphology of the cancellous bone and the density of both the cancellous and the cortical bone at the implanted zone could be used as the development criteria for better purchasing or tapping of the implant to the bone. |
| Circulation of the body fluid | Circulation of the body fluid or blood is a highly positive factor for the enhancement of osseointegration. In fact, flow of the blood at the bone-implant interface, particularly on the surface of the implant is the key for the promotion of bone formation. The blood would supply the minerals for mineralization of the new bone cells to be matured as lamellar bone. Likewise, it would flow the osteogenic cells, osteoblastic genes, proteins, cytokine genes, osteoclast cells, and other biological factors through the bone-implant interface. Today, surfaces of the implants are processed to be more hydrophilic to allow a higher rate of blood flow at the bone-implant interface. |
| Toxicity of the body fluid | Toxicity of the human body fluid would be a negative factor for the enhancement of osseointegration. One of the main reasons of the blood toxicity after implantation is the release of the toxic elements from the implant material such as arsenic, vanadium, nickel, etc. Nowadays, it is attempted to process the metallic alloys from the elements without the toxicity effect. |
| Infection of the implanted zone | Infection or growth of bacterial microorganisms at the implanted zone would be a negative factor for the enhancement of osseointegration. This parameter is a very sensitive and vital factor for successful implantation of the implant and is highly taken care by surgeons and other clinicians intra- and postoperatively. Adhesion and growth of bacterial microorganisms on the implant surface would prevent adhesion and proliferation of the osteogenic cells on the surface of the implant, thus inhibiting biological integration of the bone to the implant and after a while deducting the mechanical integration of the implant to the bone and causing loosening of the implant. |

integration was contemplated as the primary integration and biological integration of the bone through the surface or even inside the implant as secondary integration which is initiated immediately after implantation up to several months or years postimplantation. In some cases, osseointegrated implants are mainly designed and developed on the basis of bone growth in their internal holes or caging space to allow the formation of the bone tissue through the body of the implant. This concept was introduced as the endosseous concept and some of the implants which were developed considering this concept were reviewed in Section 1.5. As the final section in the first part, the expression of biological factors such as osteoblastic genes was elaborated and its effect on the growth of bone formation was highlighted.

In second part of this chapter, loosening of the implant after implantation to the bone or in contact with the bone tissue is reviewed and discussed. One of the main causes of implant loosening is wear particles which would be mostly observed in TJR due to articulation and movement between two components of the implanted prostheses at the joint. Sections 1.7 and 1.11 expressed how such generated particles could interfere with the biological conditions of the bone-implant interface, and thus reducing the bone and implant biological integrity and ultimately loosening of the implant in long-term implantation. In Section 1.8, overstimulation of mechanical signals which would be generated as the severe strain to the bone tissue was highlighted. It was observed that loading condition higher than that of natural physiological condition in health state could increase the activity of the osteoblast cells for the resorption of bone tissue at the interface and loosing of the implant might be resulted at the end. Other effective parameters on implant loosening such as synovial fluid, bone-cement interdigitation, surgical implantation, and cortical bone-implant integration were also reviewed in this chapter. Effective parameters on implant loosening are summarized in Table 1.2.

Overall, failure in osseointegration could be associated with loosening of the implant if the cause of failure is continued. In future development of orthopedic implants, such causes of the failure are advantageous to be deeply evaluated and possible solutions to prevent them are investigated. These solutions are reviewed and their effects on enhancement of long-term osseointegration are discussed in the next chapters. It is highly encouraged to contribute new development concepts for the deduction of the effect of failure in osseointegration in the implantation of the orthopedic implants without performing revision surgeries.

**Table 1.2** Effective parameters on implant loosening after implantation of the orthopedic implants

| Parameter | How to effect |
|---|---|
| Contact wear particles | As reviewed in Sections 1.7 and 1.11, release of contact wear particles would induce the expression of macrophage factors from which the giant cells might be formed due to nonresorption of the released particles. This would cause creation of granuloma at the bone-implant interface. In addition, increase in the expression of macrophage factors would increase the activity of osteoclast cells and affect the bone remodeling process at the bone-implant interface. The effect of contact wear particles is bone resorption and osteolysis by which the bone tissue at the contact surface is reduced or weakened and, ultimately, the implant would lose its integration with the bone and might be loosened or migrated (Fig. 1.2). |
| Over mechanical stimulation | Overloading conditions at the bone-implant interface would cause higher mechanical stimulation signals than that required for the formation of the new bone tissue; therefore, bone resorption is increased. For long-term implantation such as in total joint replacements, with increase in depth of bone resorption at the interface, the bone-implant integrity is reduced and it might be exceeded until loosening of the implant at the initial implantation place. |
| Failure in integrity of bone-cement bonding | Failure in integrity of the bone-cement would be one of factors that causes implant loosening in the implantation of total joint replacements. The key factor of the bone-cement integrity is interdigitation of the cement to the bone from which the shear strength at the bone-cement is increased. However, high interdigitation of the cement may cause the risk of cement particles creation and thus weakening of the bone-cement bonding. This would necessitate the consideration of optimum conditions of cement interdigitation to the bone. |
| Synovial fluid | Generally, synovial fluid has a positive effect in promoting bone formation, particularly, at the |

*Continued*

**Table 1.2** Effective parameters on implant loosening after implantation of the orthopedic implants—cont'd

| Parameter | How to effect |
|---|---|
| | joints. It has been observed that loosening of the implant would cause more production of the synovial fluid than the natural level at the affected zone which would increase the risk of brittle bone creation without effective mineralization. |
| Surgical implantation | Implantation of the implant through the bone may cause with the enhancement of severe applications of loading or thermal conditions. Steps of implantation such as drilling force and drilling rising temperature may increase the overcutting or overheating of the bone tissue from which the undesired gap might be created between the bone and the implant. If this gap could not be filled by the formation of the new bone tissue, then loosening of the implant would result. |
| Cortical bone-implant integration | Effective integration of the implant and the cortical bone could have a significant effect in preventing loosening of the implant through the trabecular bone. In fact, successful implantation to the trabecular bone is enhanced by the fit and high shear strength integration of a portion of the implant with the cortical bone. |
| Level of transferred strain to the bone | The level of loaded strain of about 1000–1500 με has been reported as the optimal condition for the bone formation at the bone-implant interface. Fewer or higher level of strain may cause resorption of the bone tissue in the long-term implantation and thus increasing the risk of implant loosening. |

## 1.14 REMIND AND LEARN

**Q1.1** What is the difference between biocompatibility and bioactivity? Explain how biocompatible material would not be sufficient for successful clinical outcomes.

**Q1.2** When was the biological reaction between the bone and the implant discovered? When the concept of osseointegration was introduced and by whom?

**Q1.3** What are the effective parameters on osseointegration? Express how each parameter influences the enhancement or deterioration of the osseointegration.

**Q1.4** What is the main development concept for the enhancement of osseointegration in implantation of the orthopedic implants?

**Q1.5** What surface parameters could influence the enhancement of the osseointegration? How the morphology and density of the bone tissue could be effective in the enhancement of osseointegration?

**Q1.6** Explain the bioactive function of the bone tissue. How this character could enhance the integrity of the bone to the implant?

**Q1.7** Express the inertness of the bioinert materials in interaction with bone. What are the consequences of the bioinert implant on osseointegration in long-term implantation?

**Q1.8** What is the advantage of inertness character of the implant material? (Explain your answer in consideration of biocompatibility aspect.) What is the effect of released particles from the implant to the bone formation process? What type of materials could have negative effects on the promotion of osseointegration?

**Q1.9** Explain how mechanical stimulation signals could be effective in promotion or deterioration of the osseointegration. How the low and high level of mechanical stimulation could be generated at the bone-implant interface?

**Q1.10** Explain why circulation of the body fluid or blood is highly crucial for the osseointegration of the orthopedic implant. What types of biological factors would be supplied by the blood for bone remodeling process?

**Q1.11** Express the effects of blood toxicity and infection at the implanted site in deterioration of the osseointegration.

**Q1.12** What are primary and secondary steps of osseointegration?

**Q1.13** Express the endosseous concept for the enhancement of osseointegration. What types of orthopedic implants would be developed considering this concept?

**Q1.14** Elaborate the gene expression. What types of noncollagenous molecules or genes are expressed in bone remodeling process? What types of mechanical and biological factors would influence the extent of gene expression and how?

**Q1.15** What are the effective parameters on implant loosening? Provide a brief explanation on how each parameter would increase the risk of implant loosening.

**Q1.16** How would wear particles influence loosening of the implant? Such failure is occurred in what type of orthopedic implants and why?

**Q1.17** What is the body reaction to reduce the inadequate effect of wear particles? Discuss whether more supplement of the body reaction material to the affected zone could be helpful to inhibit the osteolysis and implant loosening.

**Q1.18** How would over mechanical stimulation signals (or over mechanical loading conditions) cause creation of gap between the implant and the bone? Express creation of over mechanical condition in the implantation of glenoid component of total shoulder replacement.

**Q1.19** For what reason(s) the integrity of the bone and bone cement might be affected in long-term implantation? What is the current solution for the reduction or detachment of the bonding between the bone and the cement? Explain your answer in consideration of the optimal conditions.

**Q1.20** What is the role of synovial fluid for the enhancement of bone remodeling process at the body joints? How synovial fluid may have undesired effect in cases with implant loosening?

**Q1.21** How would granuloma be created after release the of the wear particles in implantation of TJR? In which type of replacement prosthesis, such biological disorder has the highest risk of occurrences?

**Q1.22** Surgical implantation procedure or instrument may have effect on loosening of the implant, discuss how.

**Q1.23** What is the effect of adequate purchasing or tapping of the implant to the cortical bone in successful implantation of the implant through the trabecular bone?

## 1.15 THINK AND CHALLENGE

**Q1.24** How circulation of the body fluid could be increased on the surface of the implant? Discuss the biomechanical and biological challenges of suggested methods.

**Q1.25** What types of methods could be utilized for the enhancement of osseointegration of the orthopedic implants after implantation.

**Q1.26** Discuss how the mechanical inertness of the current orthopedic implant could be reduced for better adoption with the flexible mechanical structural of the bone tissue.

**Q1.27** Discuss how the endosseous concept could be developed for the enhancement of orthopedic implants integration to the bone such as that happens in dental, spinal cages, and osseointegrated implants.

## REFERENCES

[1] A.I. Birkhold, et al., The influence of age on adaptive bone formation and bone resorption, Biomaterials 35 (34) (2014) 9290–9301.
[2] R. Branemark, et al., Osseointegration in skeletal reconstruction and rehabilitation: a review, J. Rehabil. Res. Dev. 38 (2) (2001) 175.
[3] K. Hagberg, et al., Osseoperception and osseointegrated prosthetic limbs, in: Psychoprosthetics, Springer, 2008, pp. 131–140.
[4] M. Klinger, et al., Proteoglycans at the bone-implant interface, Crit. Rev. Oral Biol. Med. 9 (4) (1998) 449–463.
[5] I. Nishimura, Genetic networks in osseointegration, J. Dent. Res. 92 (12_suppl) (2013) 109S–118S.
[6] T. Ogawa, I. Nishimura, Genes differentially expressed in titanium implant healing, J. Dent. Res. 85 (6) (2006) 566–570.
[7] D. Van Steenberghe, From osseointegration to osseoperception, J. Dent. Res. 79 (11) (2000) 1833–1837.
[8] F. Butz, et al., Harder and stiffer bone osseointegrated to roughened titanium, J. Dent. Res. 85 (6) (2006) 560–565.
[9] R. Jimbo, et al., Nano hydroxyapatite-coated implants improve bone nanomechanical properties, J. Dent. Res. 91 (12) (2012) 1172–1177.
[10] J. Wolff, et al., Finite element analysis of bone loss around failing implants, Mater. Des. 61 (2014) 177–184.
[11] S. Landgraeber, et al., Adiponectin attenuates osteolysis in aseptic loosening of total hip replacements, Acta Biomater. 10 (1) (2014) 384–393.
[12] J. Gallo, et al., Contributions of human tissue analysis to understanding the mechanisms of loosening and osteolysis in total hip replacement, Acta Biomater. 10 (6) (2014) 2354–2366.
[13] N. Bonnevialle, et al., Aseptic glenoid loosening or failure in total shoulder arthroplasty: revision with glenoid reimplantation, J. Shoulder Elb. Surg. 22 (6) (2013) 745–751.
[14] A.J. Wirth, R. Müller, G.H. van Lenthe, The discrete nature of trabecular bone microarchitecture affects implant stability, J. Biomech. 45 (6) (2012) 1060–1067.
[15] F. Yang, et al., Pathways of macrophage apoptosis within the interface membrane in aseptic loosening of prostheses, Biomaterials 32 (35) (2011) 9159–9167.
[16] D. Yang, D. Zhang, D.D. Arola, Fatigue of the bone/cement interface and loosening of total joint replacements, Int. J. Fatigue 32 (10) (2010) 1639–1649.
[17] J.A. Tsai, et al., Effects of synovial fluid from aseptic prosthesis loosening on collagen production in osteoblasts, Int. Orthop. 33 (3) (2009) 873–877.
[18] K.A. Stoffel, D.T. Yang, D. Arola, The influence of surface topography on wear debris generation at the cement/bone interface under cyclic loading, J. Mater. Sci. Mater. Med. 19 (5) (2008) 1935–1943.
[19] S. Landgraeber, et al., Extrinsic and intrinsic pathways of apoptosis in aseptic loosening after total hip replacement, Biomaterials 29 (24) (2008) 3444–3450.
[20] D. Arola, K. Stoffel, D. Yang, Fatigue of the cement/bone interface: the surface texture of bone and loosening, J. Biomed. Mater. Res. B Appl. Biomater. 76 (2) (2006) 287–297.

[21] P.-I. Branemark, Osseointegration and its experimental background, J. Prosthet. Dent. 50 (3) (1983) 399–410.
[22] N. Donos, et al., Gene expression profile of osseointegration of a hydrophilic compared with a hydrophobic microrough implant surface, Clin. Oral Implants Res. 22 (4) (2011) 365–372.
[23] K.M. Sinha, X. Zhou, Genetic and molecular control of osterix in skeletal formation, J. Cell. Biochem. 114 (5) (2013) 975–984.
[24] J.B. Lian, et al., Networks and hubs for the transcriptional control of osteoblastogenesis, Rev. Endocr. Metab. Disord. 7 (1-2) (2006) 1–16.
[25] C. Galli, et al., The importance of WNT pathways for bone metabolism and their regulation by implant topography, Eur. Cell. Mater. 24 (2012) 46–59.
[26] M. Monjo, et al., Correlation between molecular signals and bone bonding to titanium implants, Clin. Oral Implants Res. 24 (9) (2013) 1035–1043.
[27] T. Ogawa, I. Nishimura, Different bone integration profiles of turned and acid-etched implants associated with modulated expression of extracellular matrix genes, Int. J. Oral Maxillofac. Implants 18 (2) (2003) 200–210.
[28] O. Omar, et al., In vivo gene expression in response to anodically oxidized versus machined titanium implants, J. Biomed. Mater. Res. A 92 (4) (2010) 1552–1566.
[29] R. Jimbo, et al., Genetic responses to nanostructured calcium-phosphate-coated implants, J. Dent. Res. 90 (12) (2011) 1422–1427.
[30] K. Bougas, et al., Novel implant coating agent promotes gene expression of osteogenic markers in rats during early osseointegration, Int. J. Biomater. 2012 (2012) 579274.
[31] M. Bryington, et al., Osteoblastic and cytokine gene expression of implant-adherent cells in humans, Clin. Oral Implants Res. 25 (1) (2014) 52–58.
[32] S. Ivanovski, et al., Transcriptional profiling of osseointegration in humans, Clin. Oral Implants Res. 22 (4) (2011) 373–381.
[33] I. Wall, et al., Modified titanium surfaces promote accelerated osteogenic differentiation of mesenchymal stromal cells in vitro, Bone 45 (1) (2009) 17–26.
[34] G. Thalji, C. Gretzer, L.F. Cooper, Comparative molecular assessment of early osseointegration in implant-adherent cells, Bone 52 (1) (2013) 444–453.
[35] S. Landgraeber, et al., Over-expression of p53/BAK in aseptic loosening after total hip replacement, Biomaterials 27 (15) (2006) 3010–3020.
[36] T. Kadowaki, T. Yamauchi, Adiponectin and adiponectin receptors, Endocr. Rev. 26 (3) (2005) 439–451.
[37] T. Yamauchi, et al., Cloning of adiponectin receptors that mediate antidiabetic metabolic effects, Nature 423 (6941) (2003) 762.
[38] S.A. Antuna, et al., Glenoid revision surgery after total shoulder arthroplasty, J. Shoulder Elb. Surg. 10 (3) (2001) 217–224.
[39] E.V. Cheung, J.W. Sperling, R.H. Cofield, Revision shoulder arthroplasty for glenoid component loosening, J. Shoulder Elb. Surg. 17 (3) (2008) 371–375.
[40] R.H. Cofield, Uncemented total shoulder arthroplasty: a review, Clin. Orthop. Relat. Res. 307 (1994) 86–93.
[41] A. Deutsch, et al., Clinical results of revision shoulder arthroplasty for glenoid component loosening, J. Shoulder Elb. Surg. 16 (6) (2007) 706–716.
[42] J.S. Dines, et al., Outcomes analysis of revision total shoulder replacement, JBJS 88 (7) (2006) 1494–1500.
[43] R.J. Hawkins, P.E. Greis, P.M. Bonutti, Treatment of symptomatic glenoid loosening following unconstrained shoulder arthroplasty, Orthopedics 22 (2) (1999) 229–234.
[44] K.R. Sajadi, Y.W. Kwon, J.D. Zuckerman, Revision shoulder arthroplasty: an analysis of indications and outcomes, J. Shoulder Elb. Surg. 19 (2) (2010) 308–313.
[45] D.G. Kim, M.A. Miller, K.A. Mann, Creep dominates tensile fatigue damage of the cement–bone interface, J. Orthop. Res. 22 (3) (2004) 633–640.

[46] D.-G. Kim, M.A. Miller, K.A. Mann, A fatigue damage model for the cement–bone interface, J. Biomech. 37 (10) (2004) 1505–1512.
[47] D.D. Arola, D.T. Yang, K.A. Stoffel, The apparent volume of interdigitation: a new parameter for evaluating the influence of surface topography on mechanical interlock, J. Biomed. Mater. Res. A 58 (5) (2001) 519–524.
[48] D. Janssen, K.A. Mann, N. Verdonschot, Micro-mechanical modeling of the cement–bone interface: the effect of friction, morphology and material properties on the micromechanical response, J. Biomech. 41 (15) (2008) 3158–3163.
[49] M. Inomoto, et al., Elevated interleukin-12 in pseudosynovial fluid in patients with aseptic loosening of hip prosthesis, J. Orthop. Sci. 5 (4) (2000) 369–373.
[50] M.G. Li, K. Thorsen, K.G. Nilsson, Increased bone turnover as reflected by biochemical markers in patients with potentially unstable fixation of the tibial component, Arch. Orthop. Trauma Surg. 124 (6) (2004) 404–409.
[51] M.P. Loria, et al., Role of cytokines in gonarthrosis and knee prosthesis aseptic loosening, J. Orthop. Sci. 9 (3) (2004) 274–279.
[52] B. Nivbrant, K. Karlsson, J. Kärrholm, Cytokine levels in synovial fluid from hips with well-functioning or loose prostheses, J. Bone Joint Surg. (Br.) 81 (1) (1999) 163–166.
[53] A. Sabokbar, N. Rushton, Role of inflammatory mediators and adhesion molecules in the pathogenesis of aseptic loosening in total hip arthroplasties, J. Arthroplast. 10 (6) (1995) 810–816.
[54] I. Takei, et al., High macrophage-colony stimulating factor levels in synovial fluid of loose artificial hip joints, J. Rheumatol. 27 (4) (2000) 894–899.
[55] I. Vanhegan, et al., A financial analysis of revision hip arthroplasty, J. Bone Joint Surg. (Br.) 94 (5) (2012) 619–623.
[56] S. Patil, et al., Quality of life outcomes in revision vs primary total hip arthroplasty: a prospective cohort study, J. Arthroplast. 23 (4) (2008) 550–553.
[57] S. Kurtz, et al., Projections of primary and revision hip and knee arthroplasty in the United States from 2005 to 2030, JBJS 89 (4) (2007) 780–785.
[58] J. Chiba, et al., The characterization of cytokines in the interface tissue obtained from failed cementless total hip arthroplasty with and without femoral osteolysis, Clin. Orthop. Relat. Res. (300) (1994) 304–312.
[59] L.C. Jones, C. Frondoza, D.S. Hungerford, Immunohistochemical evaluation of interface membranes from failed cemented and uncemented acetabular components, J. Biomed. Mater. Res. A 48 (6) (1999) 889–898.
[60] S. Goodman, et al., Cellular profile and cytokine production at prosthetic interfaces, J. Bone Joint Surg. (Br.) 80 (3) (1998) 531–539.
[61] W.-C. Witzleb, et al., Neo-capsule tissue reactions in metal-on-metal hip arthroplasty, Acta Orthop. 78 (2) (2007) 211–220.
[62] M. Horiki, et al., Localization of RANKL in osteolytic tissue around a loosened joint prosthesis, J. Bone Miner. Metab. 22 (4) (2004) 346–351.
[63] S. Ito, et al., Histological analysis and biological effects of granulation tissue around loosened hip prostheses in the development of osteolysis, J. Orthop. Sci. 9 (5) (2004) 478–487.
[64] M. Sabbatini, et al., Different apoptosis modalities in periprosthetic membranes, J. Biomed. Mater. Res. A 92 (1) (2010) 175–184.
[65] T. Fujishiro, et al., Perivascular and diffuse lymphocytic inflammation are not specific for failed metal-on-metal hip implants, Clin. Orthop. Relat. Res. 469 (4) (2011) 1127–1133.
[66] G. Limbert, et al., Trabecular bone strains around a dental implant and associated micromotions—a micro-CT-based three-dimensional finite element study, J. Biomech. 43 (7) (2010) 1251–1261.

[67] A. Wirth, R. Müller, G. van Lenthe, Computational analyses of small endosseous implants in osteoporotic bone, Eur. Cell. Mater. 20 (2010) 58–71.
[68] E. Anitua, C. Carda, I. Andia, A novel drilling procedure and subsequent bone autograft preparation: a technical note, Int. J. Oral Maxillofac. Implants 22 (1) (2007) 138.
[69] O.C. Koldsland, A.A. Scheie, A.M. Aass, Prevalence of implant loss and the influence of associated factors, J. Periodontol. 80 (7) (2009) 1069–1075.
[70] H.M. Frost, A 2003 update of bone physiology and Wolff's Law for clinicians, Angle Orthod. 74 (1) (2004) 3–15.

## FURTHER READING

[71] J. Frölke, R. Leijendekkers, H. Van de Meent, Osseointegrated prosthesis for patients with an amputation, Unfallchirurg 120 (4) (2017) 293–299.
[72] M. Bahraminasab, et al., Aseptic loosening of femoral components–a review of current and future trends in materials used, Mater. Des. 42 (2012) 459–470.
[73] B. Melis, et al., Glenoid loosening and failure in anatomical total shoulder arthroplasty: is revision with a reverse shoulder arthroplasty a reliable option? J. Shoulder Elb. Surg. 21 (3) (2012) 342–349.

CHAPTER 2

# Effective Mechanical and Biological Factors on Osseointegration

## 2.1 INTRODUCTION

Bone remodeling is carried out based on the effective mechanical and biological factors that are permanently under alteration. The process of bone remodeling is extremely affected when the bone is injured and need to be healed. In some cases, it is necessitated to insert or implant the artificial components made of various types of biomaterials in contact with the bone for the treatment of bone fractures or for the replacement of degenerated bone tissue. Therefore, it is crucial to achieve a good integration of implant with adjacent or surrounding intact bone tissue. The attachment of bone cells to the implant surface followed by their stimulation to be mineralized and formed as the bone tissue at the interface of bone-implant would be important in primary and long-term integration of the implant to the bone. Mechanical response of bone cells to the loading conditions at the interface of the bone-implant would be significantly effective in integration of the implant to the bone. First, mechanical modulus mismatch between the bone and the implant forms a specific pattern of load transferring. If the modulus of the implant is higher than that of the bone tissue, major portion of the localized stress is transferred to the implant from which the physiological transferred strain for stimulation of bone cells is reduced. On the other hand, other parameters such as geometry of implant, surface condition of the implant, and corrosion resistance of implant in human body fluid are effective in osseointegration of the implant in contact with the bone. However, understanding of mechanical effects on the integration of bone tissue to the implant has been focused in recent decade to enhance the primary and long-term implantation of orthopedic implants in the treatment of bone injuries.

Osseointegration of orthopedic implants has been evaluated from various aspects. As reported in the literature [1–18], effective integration of the bone and the implant is influenced or enhanced by multiple biological and

*Osseointegration of Orthopaedic Implants*
https://doi.org/10.1016/B978-0-12-813384-2.00002-3

mechanical conditions at the interface of the bone-implant. Likewise, it has been well highlighted that the surface conditions of the implant would have a significant effect on the induction or stimulation of biological and mechanical factors for the implant osseointegration. In view of mechanobiological factors, stimulation of loading conditions at the interface of bone-implant for expression of osteoblastic marker genes is biologically crucial (the interaction of biological and mechanical factors in the bone formation is addressed as bone mechanobiology [19–27]). In order to achieve a good integration of the bone to the implant, remodeling process of the bone (bone deposition/resorption process) at the interface of the bone-implant is carried out by the effective presence of osteoblast and osteoclast cells, growth factors, proteins, and other biological factors [28–32]. In aspect of biological induction of bone remodeling, drug delivery of biphosphonate-based materials has been managed by grafting of the material to the implant surface. However, the correlation of osseointegration with optimum range of biphosphonate delivery is under investigation.

Evaluation of osseointegration ability of the implant is carried out by in vitro or in vivo assessment of the biological and mechanical factors that would influence or affect the quality of the implant osseointegration. Integration of the implant to the bone has been mechanically evaluated by pull-out or torque tests. These types of tests are conducted by the implantation of the implant to the animal bone and the range of the implant removal force or torque is measured to evaluate the bonding strength of the bone-implant after a couple of weeks or months postimplantation. Osseointegration has been also biologically evaluated by measuring the range of osteoblast- and osteoclast-related marker genes and total proteins of the integrated bone to the implant at various time points after implantation. Likewise, the activity of alkaline phosphatase (ALP) and lactate dehydrogenase (LDH) has been highlighted as evaluation factors for biological assessment of osseointegration.

In this chapter, the effective mechanical and biological factors or concepts on the enhancement or deterioration of the osseointegration are reviewed. First, the responses of the bone tissue to the mechanical factors are reviewed and then the effects of the bone mechanoresponsive on osseointegration are discussed. The deterioration of the bone formation rate due to reduction of the activity of the hormonal factors is then reviewed. Afterward, the mechanical and biological evaluation methods on the basis of in vivo implantation experiments are expressed. In the final sections of this chapter, the concept of the new/old bone contact surface with the implant and response of the cortical and trabecular bone at the peri-implant and distant bone areas are elaborated.

## 2.2 MECHANORESPONSIVE OF THE BONE

Mechanoresponsive of the cortical and trabecular bones has been assessed in the literature and investigated by radiographic scanning of the bone in human bodies to study the alteration of the bone mass and architecture in various conditions. Likewise, direct investigation has been carried out for insight study of bone alteration based on aging and mechanical loading conditions in animal bodies [33–51]. Trabecular bone has been under interest to be investigated to find out the feasible reasons of clinical-biological issues such as bone fracture, implant loosening, bone healing rate, bone fusion rate, etc., which would have mechanical-biological root causes. It has been reported that [51] vertebral body, proximal femur, and distal radius bones are in high risk of failure in people with low mineral density bones. In all the implantation of orthopedic implants, the implants are in direct contact with the trabecular and cortical bones. However, the stability of the fixation or replacement of the orthopedic implants is greatly influenced by the integration of the implant through the trabecular bone. For instance, in the fixation of screw through the bone in the implantation of trauma plating system or spinal screw-rod system, shear strength between the screw threads and trabecular bone is highly crucial to achieve a good stability of the fixation during fracture healing or bone fusion. As another example, in the replacement of the total joint replacement, the appropriate integration of the implanted prosthesis to the bone (e.g., hip femoral stem, knee tibial tray, and knee femoral) is essential to prevent the implant loosening under physiological dynamic loading conditions at the treated joint. On the other hand, based on Wolff's law, the trabecular bone is altered against the transferred loading conditions to the bone. In fact, mechanical characteristics of the bone are changing in accordance with loading regime to the bone. This would change the bone density and direction of the trabecular arrangement/direction from which the mechanical strength and modulus of the trabecular bone is directed to be strengthening in the loading direction to withstand the load. Physiological loading is considered as a type of dynamic loading. Reconstruction of the trabecular bone in microstructural scale toward the dynamic loading direction is a timely process and once the bone is strengthened in the loading direction, the strength of the bone is increased. This would influence the overall strength of the bone in other loading direction as well. Severe loading would affect softer bone tissues such as cartilage, disc, and even attached soft tissues to the bone such as muscle, tendon, and ligaments. If this severe loading condition is continued in the long term, degeneration of soft tissue might be resulted.

Through scanning electron microscope (SEM), the morphology of trabecular bone is different in various part of the body bones. It was shown that [52–54] the architecture of the trabecular bone (e.g., shape, size, and connectivity of trabeculaes) would not be similar in all trabecular bones of the body. Therefore, in order to quantify the architecture of the trabecular bone, some parameters such as bone volume fraction ($v_f$), anisotropy percentage, mean trabecular thickness, etc. were defined in histomorphometric literature [51]. Based on the variety of trabecular bone, its mechanical modulus varies from 20 to 500 MPa.

In order to discuss the relation of mechanical stimulation with mineralization and formation of the new bone tissue, the evaluation of healing process could be organized. Hoerth et al. [40] investigated healing of the created osteotomy in rat with different gaps. They considered the gap of 1 mm as the noncritical condition and 5 mm as the critical condition at which the bone healing might not be effectively achieved. In their study, healing of 1 mm gap was observed while the bone formation of the 5 mm gap was not achieved. It was also obtained that the magnitude of transferred loading conditions at the gap zone would be significantly higher in 1 mm gap, which would influence the differentiation and proliferation of osteogenic cells and ultimately the formation and ossification of soft callus woven fibrous followed by mineralization and formation of cortical bone tissue at the gap site.

Immobilization of the bones would result in unloading of the bone tissue from which the required mechanical stimulation for induction of osteogenic cells would not be created to facilitate differentiation and formation of the lamellar bone. Thus, the volume or density of the bone tissue is gradually reduced. Fig. 2.1 illustrates the increase of cortical and trabecular sections of the murine tibia bone under compression force compared to unloaded control case. The summary of mechanoresponsive aspects of osseointegration is given in Table 2.1.

## 2.3 EFFECT OF HORMONAL FACTORS IN MECHANORESPONSIVE OF THE BONE

In addition to the bone mechanical response, alteration of hormonal, genetic, and nutritional factors would cause changes in mass and architecture of the bone. Mechanoresponsiveness of the bone tissue to the mechanical

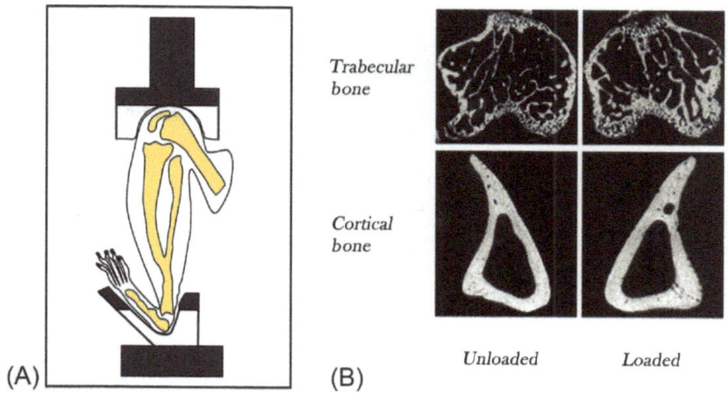


**Fig. 2.1** Bone response to mechanical loading. (A) Mechanical setup for loading of compression force to the murine tibia bone. (B) Strengthening of the trabecular and cortical bone tissues in the loading direction in loading case compared to unloading control case. *(Courtesy of L.B. Meakin, J.S. Price, L.E. Lanyon, The contribution of experimental in vivo models to understanding the mechanisms of adaptation to mechanical loading in bone, Front. Endocrinol. 5 (2014) article 154.)*

**Table 2.1** Mechanoresponsive aspects of osseointegration

| Relevant aspect | Description |
|---|---|
| Integration of the implant through the trabecular bone | Response of the bone to the implanted prosthesis would be a complex biological-mechanical scenario. The response of the trabecular bone to such condition is significantly higher than that of the cortical bone. On the other hand, the trabecular bone could be adopted with the implant in lesser time than the adoption of the cortical bone with the prosthesis or the implant. The initial response of the bone is mechanically achieved by interlocking of the implant through the bone (particularly the cortical bone) for primary stability. By transferring and distribution of the natural physiological loading conditions to the bone-implant interface, the biological response of the osteogenic cells is activated and thus bone deposition or formation is processed on the surface of the implant. Adequate response of the preimplanted |

*Continued*

**Table 2.1** Mechanoresponsive aspects of osseointegration—cont'd

| Relevant aspect | Description |
|---|---|
| | bone tissue to the prosthesis or the implant at the initial postimplantation period would be critical for primary stability of the implant in place. |
| Response to physiological dynamic loading conditions | Response of the bone to the physiological dynamic loading conditions is biologically essential for the long-term implantation of the implant to the bone. Due to direct proportion of the bone generation or the deposition to conducted mechanical stimulation signals, the continuous processing of the bone formation is governed by the continues conduction or transferring of the mechanical signals to the bone cellular matrix. Therefore, creation of cyclic or dynamic physiological loading conditions (which is obtained by movement and load bearing of the bone tissue) to the bone tissue would be a major concern in maintaining of the bone in healthy condition or in the treatment of the bone injuries through internal implantation. |
| Strengthening of trabecular bone modulus through the loading direction | Response of the trabecular bone to the conducted loading conditions would be recognized as one of the interesting mechanobiological concepts of bone biology. Due to higher porosity of the trabecular bone compared to cortical section of the bone structure, the extent of mechanical stress on the porous structure of the trabecular bone is significant. The response of the trabecular bone to counteract the loading effect is to be strengthened in the loading direction. This would increase the overall mechanical modulus of the bone structure in the loading direction. Such behavior is contemplated as the smart reaction or response to the surrounding loading conditions, which would be utilized for the development of smart materials |

**Table 2.1** Mechanoresponsive aspects of osseointegration—cont'd

| Relevant aspect | Description |
|---|---|
| | nowadays. Although, the trabecular bone has adopted its structure to be aligned with the loading extent and direction, however, the trabecular bone in all parts of the body skeletal has its specific natural structuring. Therefore, it is beneficial or even essential to conduct the loading condition in the natural direction of the trabecular bone. Strengthening of the bone in an unnatural organism may cause growth of the bone in an irregular anatomical shaping from which the surrounded soft tissues might be irritated. |
| Gap between the bone fragments or bone and implant | Response of the bone tissue to be connected or healed with other bone tissue is essentially challenging. In fact, filling of the gap between two bone fragments would necessitate the effective response of the bone tissue to create a cartilaginous fibrous tissue (as the connective bridge between the bone fragments) would be considered as one of the mechanical responsive of the bone. Creation of such connective bridge would be achieved with the existence of relative micromovement of the bone fragments in a small length of the gap (lesser than 1 mm). This would allow load transferring at the gap zone and growth of each bone fragment through the other bone fragment. Such mechanical-biological pattern could be processed between the bone and the implant for the enhancement of the implant osseointegration if the implant is coated or structured with osseoconductive layer for growth the of the bone tissue on the implant. |

*Continued*

**Table 2.1** Mechanoresponsive aspects of osseointegration—cont'd

| Relevant aspect | Description |
|---|---|
| Immobilization of the bone | Response of the bone to the mechanical loading has been described in the previous aspect of the mechanorespone. Conversely, with immobilization of the bone, converse processing of the bone formation is exceeded as bone resorption through which the bone minerals are resorbed and thus weakening of the bone structure. Such behavior would be contemplated as the smart reaction or response of the bone to counteract the unloading condition of the bone cellular matrix. Such bio-smart response would be existed in the body tissues for resorption of the affected or dead cells and generation of new mature cells for growth of the body tissues. However, all this remodeling process is carried out with continues mobilization of the body skeletal system and immobilization would cause activity of the resorption factors. |

factor/loading has been clearly observed in athletes who received high-impact and high-velocity loading conditions (e.g., tennis players whose playing arm has a higher bone mass and diameter than the nonplaying arm). Although mechanoresponsive ability of the bone tissue could encourage exercises and mechanical loading for increasing the bone density, however, bone remodeling or osteogenesis is remarkably affected by aging. It means that other than mechanical factors that would enhance the bone remodeling and the bone formation, hormonal changes would also influence the quality of bone remodeling. In women, the ability of bone formation is reduced after the menopause. Bassey et al. [56] showed that in premenopausal women, the bone mineral density (BMD) is increased by 2.8% after 5 months high-impact exercise compared to no increase in BMD of postmenopausal women with similar exercise conditions. In general, it was observed that the trabecular bone density is increased because of high-impact and high-velocity exercises in prepubescent girls [57] and premenopausal women [58–61]. However, no increase or modest increase of BMD was reported for postmenopausal women after similar exercises [62–64]. Based on the investigation of Riggs et al. (reported in 2008), the trabecular bone loss observed in women was 37% and in men was 42% for those above 50 years of age. This observation was 6% and 15% for the loss of cortical bone for women and men, respectively. Alteration in circulating of the hormones would result to lowering the productivity of the basal cells from which the density of the lining osteoblast cells is reduced [43–55], and therefore, higher level of mechanical signals is needed for stimulation of the natural bone remodeling process. However, higher level of mechanical stimulation might not be created by aging and thus the osteoblast or osteogenic cells would not be induced to be formed as osteocyte and ultimately matured to bone cellular matrix and increase of the bone mass [65–67]. Reduction in skeletal loading and muscle mass would reduce mechanical inductivity of the bone cells, however, they are not the initial causes of bone mass deterioration [67].

## 2.4 REDUCTION OF BONE MECHANORESPONSIVENESS BY AGING

Mechanoresponsiveness of the trabecular bone is reduced in adulthood compared to young age. In young age, in response to the mechanical loading, significant increase of the bone volume fraction and the thickening of trabeculae was observed in an experimental study by Willie et al. [51]. Likewise, mechanical response of the bone in adult age has been observed with longer time compared to young age. This would show a reduction of the bone mechanoresponsiveness with increasing age. In order to discuss why the bone response in adult age is lower than young age, various type of experimental investigations have been carried out. In such investigations, the level of strain for induction of the bone formation has not been precisely quantified. It is, therefore, in order to perform in vivo study to investigate the response of bone to the mechanical loading, it is important to organize the loading conditions in an adequate level for induction of the osteogenic cells for the formation of new bone tissue and increase of detectable bone volume fraction. Somerville et al. [68] achieved that the bone elastic modulus and moment of inertia at cross section of the bone is increased in adult age. This would cause a reduction of the generated strain to the adult cortical and trabecular bone at similar loading levels. In addition, the decrease of bone density in nonloading limb has been observed in adult age, while it remains constant in nonloading limb of the young age for a period of time.

The natural human skeleton may not be maintained with aging through the effective bone remodeling process (formation and resorption). Birkhold et al. [33] investigated the effect of mechanical loading on the bone remodeling process. They managed their study on mice and studied changes of bone formation and resorption using longitudinal in vivo micro computed tomography (CT) and focused on the involvement level of osteoblasts and the rate of bone formation. For complementary study of the bone remodeling activity, they have also studied the level of active osteoclasts and penetration depth of osteoclasts to the bone tissue for resorption process. In Birkhold's research, formation and resorption rate has been related to the activity of the osteoblast and osteoclast cells which was reported for young postpubescent, adult, and old mices with age of 10, 26, and 28 weeks, respectively. The activity of osteoblasts and the rate of the bone formation were naturally effective when compared to lowering the activity of osteoblast by aging. Likewise, the level of active osteoclasts and penetration depth of osteoclasts was observed in natural order in young age and higher activity of osteoclasts for resorption of the bone has been observed by aging. The summary of bone mechanoresponsiveness aspects in young age and by aging is shown in Table 2.2.

## 2.5 EFFECT OF BISPHOSPHONATE ON IMPLANT OSSEOINTEGRATION

The main method to increase the osseointegration of bioinert materials is the deposition of a bioactive layer on the surface of the material. Calcium-phosphate-based bioceramics was extensively used to be coated on the surface of the implant. It was reported that hydroxyapatite (HA) coating enhances the secondary integration of the implant by which stability and shear strength at the bone-implant interface is increased. It would also promote mineralization of the new osteocyte cells to be formed as bone extracellular matrix [69]. Furthermore, HA coating could increase the fatigue strength of the implant at the coated area due to the enhancement of surface molecular bonding [70]. However, the main concern of HA coating is the bonding strength between the implant and coating layer in long-term implantation. In fact deposition of HA layer to the implant is the major limitation of coating layer integration with the implant [71]. Although HA and bone tissue could establish a strong bonding, the bonding between the implant and the HA-coated layer might be loosen postoperatively.

**Table 2.2** Bone mechanoresponsiveness in young age and by aging

| Response type | Young age | By aging |
|---|---|---|
| Bone volume fraction | Increase of the bone volume fraction and thickening of trabeculae | Constant or reduction of the bone volume fraction by aging |
| Time to response | Mechanical response of the bone to the physiological loading conditions is in shorter time in young age compared to adult age | |
| Bone mechanical characteristics | Mechanical characteristics of the bone such as elastic modulus and geometrical characteristics of the bone such as moment of inertia at bone cross section are increased in adult age which would cause reduction of generated strain to the adult's cortical and trabecular bone at similar loading levels. | |
| Bone density in nonloading limb | Constant density of the bone in nonloading limb of the young age for a period of time. | Decrease of the bone density in nonloading limb has been observed in adult age. |
| Bone remodeling process | Natural human skeleton is constant through effective processing of the bone remodeling process (bone formation and resorption) | Natural human skeleton may not be maintained with aging through effective bone remodeling process (bone formation and resorption) |
| Activity of osteoblast and osteclast | Involvement of osteoblasts and the rate of bone formation are naturally effective. Likewise, the level of active osteoclasts and penetration depth of osteoclasts was observed in a natural order. | Involvement of osteoblasts and the rate of the bone formation has deterioratedwith aging. Likewise, the level of active osteoclasts and penetration depth of osteoclasts was observed higher than the natural conditions in young age. |

In order to enhance the effectiveness of the coating layer for long-term implantation, grafting of biphosphonate-based materials [e.g., pamidronate (PAM), ibandronate (IBA), and zoledronate] to the HA-coating layer was investigated [1] (delivery of zoledronate has been found with a better osseointegration effect than PAM and IBA [72, 73]). Bisphosphonate has been utilized as the drug to treat the osteoporosis and has been addressed as the local delivery factor for the enhancement of osseointegration [6, 73–77]. Positive effect of bisphosphonate to decrease primary osteolysis of the peri-implant bone tissue was observed [1] from which the risk of aseptic loosening was reduced in total hip arthroplasty. Peter et al. [1] achieved that the amount of bisphosphonate on contact interface of the bone-implant (which was grafted to the implant surface) should be optimal. They found that the pull-out force of inserted implant to the rabbit femur is maximal for the grafted bisphosphonate of 0.2 mm while the force for 2 mg is even lower than the implant without grafting bisphosphonate. This would challenge the use of bisphosphonate for the enhancement of osseointegration.

## 2.6 BIOLOGICAL-MECHANICAL EVALUATION OF OSSEOINTEGRATION THROUGH PULL-OUT TEST

Manjo et al. [4] investigated the osseointegration of the titanium implant to characterize the biological conditions of attached bone tissue to the implant. They intended to evaluate the level of osseointegration in relation with the range/existence of LDH and ALP markers, protein, and bone-related gene expression markers. They utilized pull-out test to examine the strength of the bone-implant integration to express the effect of mentioned biological factors on osseointegration. In fact, these biological factors are active in the process of the bone remodeling (bone resorption/deposition). After pulling out the implant, the relation of each factor to the osseointegration could be evaluated. The remaining attached bone tissue to the pulled-out implant is used for evaluation of the total protein (by quantification using protein assay kit) and DNA/RNA (by using spectrophotometer). Likewise, the remained tissue fluid at the implanted hole is taken (could be taken by filter paper for absorbing of the tissue fluid) for evaluation of ALP and LDH activities. Total protein could also be evaluated from the tissue fluid for better and accurate examination of the activity level of ALP and LDH. Based on the observations of Manjo et al. [4], the pull-out test would be a good method for evaluation of bone attachment strength to the implant as stated by other investigators [78, 79]. However, they highlighted using of the coin-shape implant rather than the thread (like screw) feature implant. This preferred implant shape is to minimize the shear force between the bone and the implant which is resulted by mechanical engagement of the implant to the bone (with purchasing of the implant threads to the bone, mechanical engagement of the bone and the implant is resulted, from which the deposited bone tissue to the implant could be partially remained after removing from the implanted hole, whereas the biological factors might not be remained on the implant during removing from the bone). On the other hand, in majority of the studies, surface conditions of the implant are investigated to find out the strength of the attached bone to the implant. In such studies, surface roughness, bioactive coating, mechanical engagement, and surface treatments are studied by performing of pull-out or removal torque tests at which the interest is only the shear strength of the bone-implant integration and evaluation of biological factors at the bone-implant interface is not required (Fig. 2.2 would show the schematic illustration of the pull-out test along with other types of mechanical testing methods for the evaluation of the bonding strength between the bone and the implant). In fact, minimal bone-implant engagement would allow easier pulling out of the implant and remaining of the biological factors on the surface of the implant for further evaluation. As a conclusion in this part, pull-out test using the coin-shape implant (or implant with other smooth shape and easy to pullout) is preferred for biological study of the bone-implant interface and pull-out/removal torque test using the real implant shape with various surface conditions is preferred for the evaluation of shear strength between the bone and the implant.

## 2.7 BIOLOGICAL EVALUATION OF OSSEOINTEGRATION THROUGH DETECTION OF BIOLOGICAL FACTORS

Bone remodeling could be contemplated as a biological process at which osteoblast cells are constructing the bone cellular matrix. In this process, once the osteoblastic genes are expressed, the collagenous fibers (fibrils) are synthesized in the surrounding of osteoblast cells. Then, these collagen fibrils are mineralized by organic constituents such as calcium phosphate. The steps of bone remodeling are organized through existed proteins in the tissue fluid at the processed zone. Therefore, the reduction of the protein level at the tissue fluid of the bone-implant interface could show the progressing of the bone remodeling [4, 79]. Likewise, the activity of ALP would show that mineralization process in bone remodeling is ongoing which

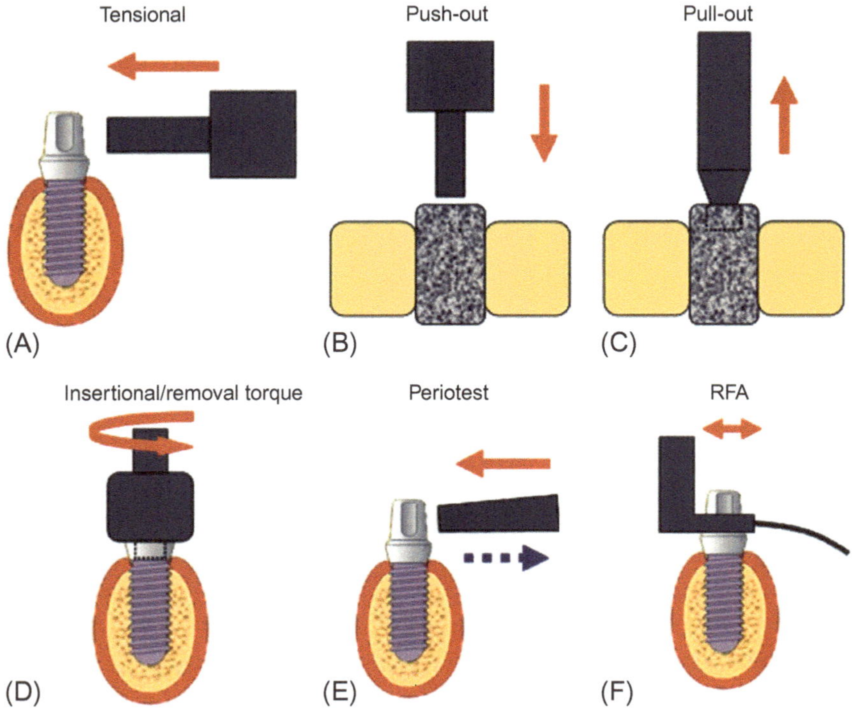


**Fig. 2.2** Schematic illustration of the pull-out test along with other types of mechanical testing methods for the evaluation of bonding strength between the bone and the implant (A) tensional test, (B) push-out test, (C) pull-out test, (D) removal torque test, (E) periotest, and (F) RFA test. Pull-out and removal torque tests are normally utilized to examine the enhancement of the osseointegration in the development of orthopedic implants. *(Courtesy of P.-C. Chang, N.P. Lang, W.V. Giannobile, Evaluation of functional dynamics during osseointegration and regeneration associated with oral implants, Clin. Oral Implants Res. 21 (2010) 1–12 and permission from the Clinical Oral Implant Research.)*

results in the formation of mature osteocyte cells. Based on the reported observations [4, 80], reduction in the level of ALP in the fluid at the integrated tissue to the implant, demonstrates the steady state of mineralization process from which a better integration of the bone to the implant could be concluded. It has been achieved that [4] decrease in the level of LDH in the tissue fluid could reveal lower level of inflammation. Thereby, it could be affirmed that the bone formation at the interface of the bone-implant has been passed at the early stage of the bone formation process to build the cartilaginous tissue between the bone and the implant (the same scenario is occurred in healing of bone fragments at the fracture gap). Low range of LDH activity would be the sign of a steady-state bone-remodeling process at the interface of the bone-implant. As found by Manjo et al. [4], total protein in the tissue fluid and the integrated tissue to the implant, after implant removal, was reduced at 8 weeks postimplantation compared to 4 weeks which indicated the normal biological condition of the bone remodeling at the interface. A higher amount of osteoblastic marker gene (e.g., osteocalcin, coll-I, runx-2, and IGF-I) was detected in the integrated tissue to the implant at 8 weeks implantation compared to 4 weeks. In contrast with the factors; total protein, LDH, and ALP, increase of osteoblastic genes correlates with the increment of osseointegration.

Similar to osteoblastic marker genes, osteoclast-related markers "TRAP" and "CalcR" have been detected higher in the integrated bone tissue to the implant in 8 weeks postimplantation compared to 4 weeks [4, 81]. This would be other evidence that bone remodeling is ongoing at the bone interface. In fact, osteoblast marker genes are indicative of deposition or generation of new bone tissues while osteoclast markers show resorption of degenerated bone tissues. Among the osteoblastic genes, the expression of osteocalcin and collagen type I genes has been addressed to have a significant effect on the bone generation during integration of the bone to the implant. As highlighted in the literature [82–84], osteocalcin expression is for non-collagenous protein of the bone and its high level in the integrated tissue would show higher formation of mineralized osteoblast cells (osteocyte cells). Likewise, higher presence of collagen-I indicates the formation of lamellar bone as it would constitute about 90% of the bone matrix. It was achieved that mechanical strength of bone is correlated with expression of collagen gene in strengthening of woven bone to be formed as lamellar bone (lamellar bone is the bone with full formation of organic constitutes into the bone matrix). Table 2.3 would describe the biological factors that are investigated for evaluation of implant osseointegration.

## 2.8 RATIO OF NEW/OLD BONE CONTACT SURFACE WITH IMPLANT

When the implant is inserted to the bone, the parent or old bone, which has been crushed by the implant, could not be integrated with the implant. In fact, the old bone is replaced by the new bone tissue. The increase of ratio of the new bone contact surface over the old bone contact surface with the implant at the interface of the bone-implant would indicate the progress of osseointegration [7, 10, 12, 85, 86]. The old bone is resorbed and the new bone is generated through the surface of the implant to provide

**Table 2.3** Biological factors for evaluation of the implant osseointegration

| Biological factor | How to examine the osseointegration |
|---|---|
| Protein | Low level of protein at the bone-implant interface could demonstrate the high level of the bone remodeling process either bone formation regeneration and bone resorption. |
| Alkaline phosphatase (ALP) marker | With detection of the ALP at the interface of the bone-implant, mineralization of the osteocyte cells could be concluded. |
| Reduction of ALP | With reduction in the level of ALP, the steady state of mineralization process is resulted. This would be a good sign in formation of the bone on the surface of the implant from which the progression of the osseointegration is concluded. |
| Lactate dehydrogenase (LDH) | Decrease in level of LDH in the tissue fluid could reveal lower level of inflammation. Thereby, it could be affirmed that the bone formation at the interface of the bone-implant has been passed the early stage to build the cartilaginous tissue between the bone and the implant (the same scenario is occurred in healing of the bone fragments at the fracture gap). Low range of LDH activity would be the sign of steady-state bone-remodeling process at the interface of the bone-implant. |
| Total Protein | Reduction of total protein after about 8 weeks postimplantation would be a positive sign for creation of new bone matrix at the interface of the bone-implant. |
| Osteoblastic marker genes | Higher amount of osteoblastic marker genes after about 8 weeks postimplantation (e.g., osteocalcin, coll-I, runx-2, IGF-I) correlates with increment of osseointegration. |
| Osteoclast-related markers | Osteoclast-related markers "TRAP" and "CalcR" after about 8 weeks postimplantation is the other evidence of effective bone remodeling process at the bone-interface. |
| Osteocalcin and collagen type I genes | Expression of osteocalcin and collagen type I genes would show higher formation of mineralized osteoblast cells (osteocyte cells). Likewise, higher presence of collagen-I indicates formation of lamellar bone as it would constitute about 90% of the bone matrix. It was achieved that mechanical strength of the bone is correlated with expression of collagen gene in strengthening of woven bone to be formed as lamellar bone |

integration of the bone and implant. It was observed that the implant geometry has higher effect on the speed of osseointegration than the density or morphology of the bone as cortical or trabecular bone tissue. The rate of the formation of new bone tissue is influenced by various factors including implant geometry and surface conditions, bone morphology, and loading conditions. Beutel et al. [10] investigated various types of screws with different thread profiles. The ratio of the new bone-implant contact has been achieved in different values for each thread profile. They have highlighted the importance of the contact surface between the bone-implant in the enhancement of implant osseointegration. Formation of the woven bone and its remodeling to the lamellar bone at the bone-implant interface (Fig. 2.3) would be the target of osseointegration which might be progressed from bone to the implant or from the implant to the bone.

In one type, the osteogenesis cells may be adhered, trapped, or absorbed to the implant surface and thus the new bone tissue deposition is initiated and progressed from the implant surface through the bone tissue. In other type, the formation of the bone might be initiated from the distant osteogenesis cells and then the woven bone is gradually deposited through the

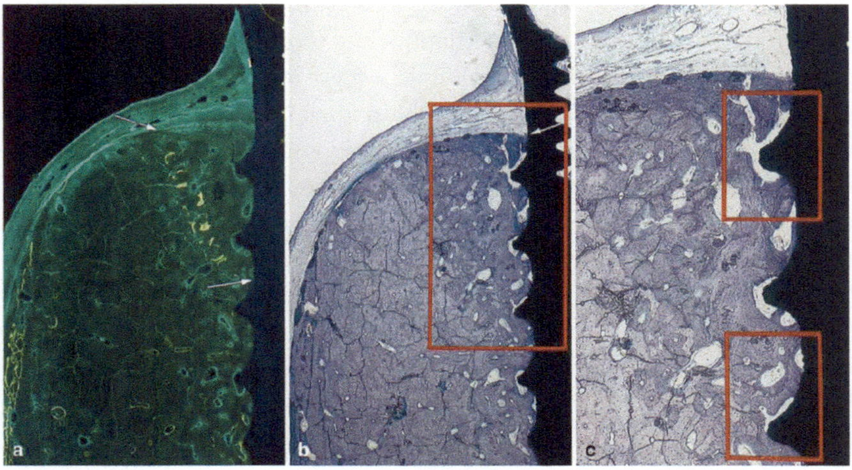


**Fig. 2.3** Bone-implant interface. (A) Fluorescence light (magnification ×25). The bone defect is demonstrated in yellow color. Adherence of the new lamellar bone to the implant surface; the black color is the implant, the dark pink color is illustrating the bone tissue, and the light color at the interface of the bone and implant is connective tissue; (B) magnification ×25, (C) magnification ×50. *(Courtesy of L.G. Persson, et al., Re-osseointegration after treatment of peri-implantitis at different implant surfaces, Clin. Oral Implants Res. 12(6) (2001) 595–603 and permission from the Wiley Online Library.)*

implant surface and remodeled to the lamellar bone. The latter type would be more theorized for the effective enhancement of the osseointegration [10, 87] and is greatly under interest for the improvement of implant surface conditions to increase the extent of absorption or adhesion of the osteogenesis cells, mesenchymal stem cells, and other biological factors, precursors, genes involved in bone remodeling process.

## 2.9 EVALUATION OF BONE REMODELING AFTER INSERTION OF THE IMPLANT

After insertion of the implant through the bone, the response of the cortical and trabecular bone to be integrated with the implant would be different. In fact, bone remodeling process is significantly affected at the bone tissue near to the implant (peri-implant). On one hand, the distribution and extent of the mechanical loading is highly altered by the implantation of a new device through the bone, from which the mechanical stimulation signals for the bone formation might not be naturally generated. On the other hand, flow of blood and the presence of biological factors at the implanted zone is greatly affected, by which the bone formation might be significantly reduced. In order to evaluate the effect of implantation of a medical implant through the bone tissue, the bone remodeling (bone resorption and formation) at the bone tissue near to the implant (peri-implant) and distant to the implant was studied by Li et al. [12]. In their investigation, cylindrical implants from aluminum composite were inserted in caudal vertebrae (CV6) of mice as schematically displayed in Figs. 2.4B and 2.5A. As illustrated in Fig. 2.4A, the cortical thickness (Ct.Th) is increased in the peri-implant bone tissue compared to distant tissue. This would show the higher rate of the bone formation in the cortical bone at the implanted area due to the overloading of the natural physiological loading that is normally transferred to the cortical bone. Due to high gripping strength of the implant to the cortical bone, greater level of mechanical strain might be transferred to the peri-implant cortical bone and thus, it could increase the accumulation of the blood and impregnation of the biological factors at the interface of the bone and the implant. Ultimately, with the presence of excessive mechanical stimulation signals and biological factors at the peri-implant cortical bone, the process of the bone formation is accelerated. It is interesting to highlight that the level of mechanical strain at the peri-implant zone is stabilized after a couple of weeks, and subsequently the Ct.Th is constant as that observed in distant cortical bone which has mild increase of cortical bone thickness after

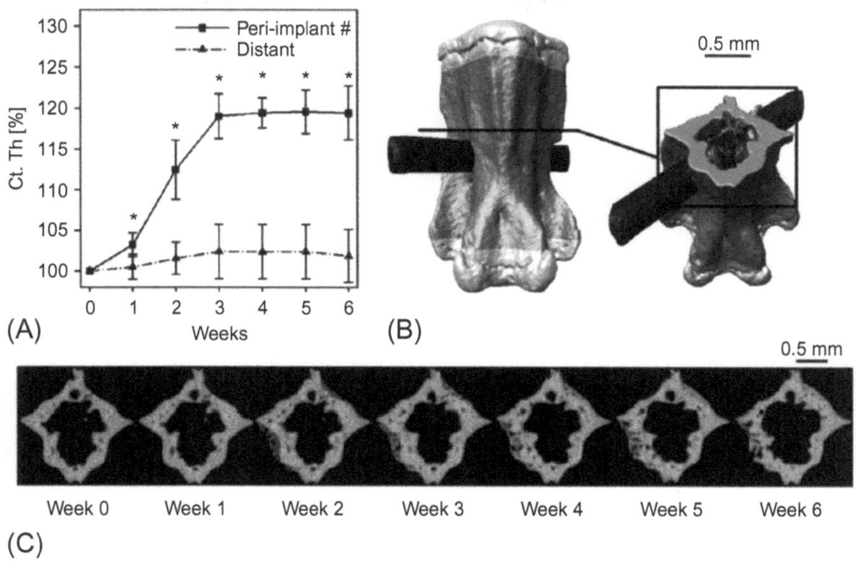


**Fig. 2.4** The effect of implantation on cortical bone, (A) the cortical thickness alteration up to 6 weeks postimplantation, (B) illustration of the implant insertion through the caudal vertebrae of the mice and clarification of the peri-implant and distant cortical bone areas, (C) cross-sectional illustration of the cortical thickness in peri-implant and distant areas up to 6 weeks postimplantation. *(Courtesy of Z. Li, et al., In vivo monitoring of bone architecture and remodeling after implant insertion: the different responses of cortical and trabecular bone, Bone 81 (2015) 468–477 and permission from Elsevier.)*

implantation (Fig. 2.4A and C). In converse to what has been observed in the cortical bone, the trend of the bone formation at the peri-implant trabecular bone is remarkably lower than the distant trabecular bone tissues. The bone formation or bone volume fraction (BV/TV) in both peri and distant bone areas is reduced during 6 weeks implantation (Fig. 2.5) and the rate of this reduction might be reduced after 6 weeks. These results would reveal the high level of trabecular bone deterioration or weakening after implantation. The reason might be the significant lowering of the transferred mechanical strain to the trabecular bone which would occur due to the poor gripping or tapping of the trabecular bone to the implant. It could be concluded that the enhancement of the implant osseointegration would be greatly investigated on increase of gripping or shear strength between the implant and the trabecular bone (e.g., by featuring of the threads to the implant surface such as screws or roughening of the implant surface such as femoral component of the total knee replacement). Other surface

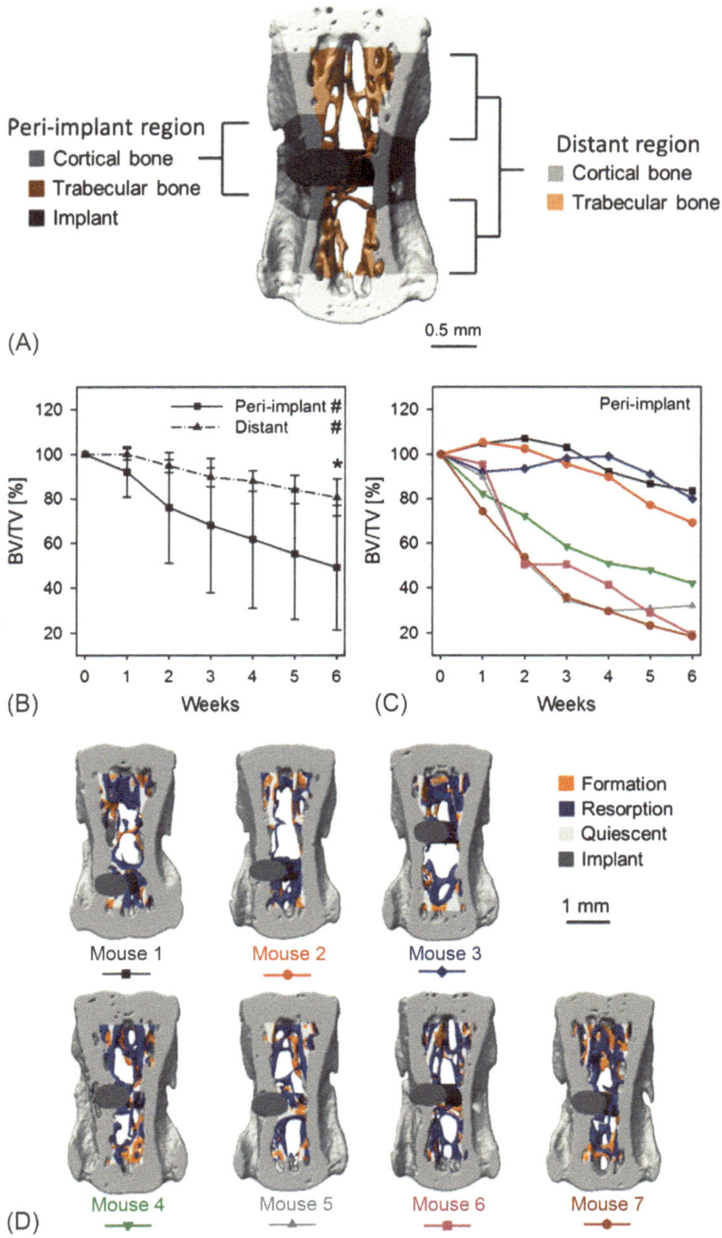


**Fig. 2.5** The effect of implantation on the trabecular bone, (A) illustration of the implant insertion through the caudal vertebrae of the mice and clarification of the peri-implant and distant trabecular bone areas, (B) alteration of the bone volume fraction at the peri-implant and distant trabecular bone areas up to 6 weeks postimplantation, (C) alteration of the bone volume fraction at various location of implantation with different thickness of cortical bone and density of porous trabecular bone , and (D) illustration of the bone formation and resorption at various location of implantation in each animal. *(Courtesy of Z. Li, et al., In vivo monitoring of bone architecture and remodeling after implant insertion: the different responses of cortical and trabecular bone, Bone 81 (2015) 468–477 and permission from Elsevier.)*

processing methods such as bioactive coating and osseoconductive porous structuring are currently developed to enhance the adhesion of the biological factors to the implant surface and accelerate the bone formation at the interface of trabecular bone and the implant surface or at the peri-implant bone [12].

## 2.10 SUMMARY

Due to bioactive nature and mechanical structuring of the bone tissue, there would be various biological and mechanical factors which would influence the bone remodeling process at the interface of the bone and the implant and thus implant osseointegration in the long-term implantation. In this chapter, these factors were reviewed generally and constructive discussions were provided for a better understanding or challenging of the expressed concepts. Initially, the mechanical response of the bone was elaborated as the mechanoresponsive behaviors of the bone against of physiological loading conditions. As reviewed, the mechanoresponsive behavior of the bone is greatly affected by the implantation of an artificial device or implant through the bone. As highlighted in Table 2.1, the mechanoresponsive aspects of the bone such as integration of the implant through the trabecular bone, response to physiological dynamic loading conditions, strengthening of trabecular bone modulus through the loading direction, gap between the bone fragments or the bone and the implant, and immobilization of the bone are crucial to be investigated for the better enhancement of the implant osseointegration. The mechanoresponsiveness of the bone would be deducted by aging due to the alteration of the hormonal factors to induce the effective biological factors in processing of the bone remodeling as reviewed and discussed in Sections 2.3 and 2.4. The effect of bisphosphonate on the enhancement of bone formation has been expressed in Section 2.5 as the effective factor to accelerate the implant osseointegration. However, care must be taken that the level of bisphosphonate must be in the optimal range to have efficacy in promotion of the bone-implant integration. In order to assess the improvement of osseointegration in the development stage, mechanical and biological evaluation tests would be beneficial. In this regard, the processed sample is implanted through the animal bone and after a couple of weeks postimplantation (might be at 4, 6, 8, 12, or 16 weeks) pushout or remove torque along with analysis of the peri-implant blood and implant-integrated tissue are carried out to evaluate the osseointegration of the implant at in-vivo conditions. Higher amount of force or torque to remove the implant from the bone would show the greater osseointegration of the implant. Likewise, with quantifying the activity level of the biological factors (e.g., protein, ALP, LDH, osteoblastic marker genes, osteoclast-related markers, osteocalcin gene, and collagen type I gene), the progression of the bone formation at the interface of the bone and the implant could be predicted from which the effect of applied process on the enhancement of the osseointegration would be studied. The ratio of the new and old bone contact surface with the implant was expressed as the biological evaluation criterion to examine the effectiveness of the utilized processing method for enhancement of the osseointegration. Higher rate of the new bone contact with the implant would reveal greater activity of the bone formation or generation at the bone-implant interface. As the final section of this chapter, the effect of implantation on the bone remodeling process of the cortical and the trabecular bone tissues was reviewed according to the published investigation carried out by Li et al. [12] and feasible reasons were discussed for better expressing of the achieved results. It was observed that the bone formation is increased at the peri-implant cortical bone and remarkably decreased at the peri-implant trabecular bone. The reviewed effective mechanical and biological factors in the implant osseointegration could be used for the establishment of various types of evaluation methods to assess the improvement of osseointegration in the development stage. In the following chapters, these improvement methods are reviewed and discussed in the enhancement of osseointegration in various types of orthopedic implants such as trauma fixation implants, joint replacements, spinal fusion implants, and tissue scaffolds.

## 2.11 REMIND AND LEARN

**Q2.1** Express how the stability of the implants might be affected. What is the role of the trabecular bone in the enhancement of the implant stability? Provide examples in your answer.

**Q2.2** What characteristics of the trabecular bone might be altered for bearing of the physiological loading?

**Q2.3** What is the range of mechanical modulus of the trabecular bone? Explain why the mechanical modulus is different in various parts of the body.

**Q2.4** What is the relation between the mechanical stimulation and mineralization and formation of the new bone tissue? Express your answer with consideration of the healing process at the fracture gaps between bone fragments. Discuss how the gap between the bone and the implant could affect the implant osseointegration.

**Q2.5** What is the effect of the body immobilization on the bone formation? Consider Fig. 2.1 for a better answering to this question.

**Q2.6** What are the mechanoresponsive aspects of osseointegration? Provide a brief elaboration for each aspect.

**Q2.7** What is the effect of hormonal factors in mechanoresponsiveness of the bone? How the density of lining osteoblast cells would be reduced with aging? Express this effect in premenopausal and postmenopausal women.

**Q2.8** What are the bone mechanoresponses to the physiological loading conditions? Express this response in young age and its deterioration with aging.

**Q2.9** What is the effect of biphosphonate-based materials on osseointegration? What types of biphosphonate are currently utilized and which type has greater effect on the implant osseointegration? Discuss why the optimal range of biphosphonate is beneficial to be delivered or released at the bone-implant interface.

**Q2.10** What types of mechanical testing method might be utilized for the evaluation of implant osseointegration?

**Q2.11** What types of biological factors might be detected for the evaluation of osseointegration progression at the bone-implant interface? How this detection would be carried out? Provide a brief explanation on how each type of biological factor would examine the progression or quality of the osseointegration.

**Q2.12** What are the meaning of the new and the old bone tissue in contact surface with the implant? How the ratio of the new/old bone contact surfaces with the implant could be used for evaluation of the osseointegration?

**Q2.13** What are the effective factors on the rate of new bone formation at the bone-implant interface?

**Q2.14** Express the remodeling of the new bone tissue at the bone-implant interface and how it could be progressed. Which type of the new bone remodeling has been theorized and how the implant osseointegration could be enhanced by consideration of such theories?

**Q2.15** Discuss the effect of implantation of an artificial implant on the bone remodeling behavior of the cortical and trabecular bone at near and distant areas of the implant. Provide feasible reasons for your answers.

## 2.12 THINK AND CHALLENGE

**Q2.16** What types of in-vivo evaluation methods could be utilized for examination of the long-term osseointegration after a couple of months or even years of implantation?

**Q2.17** What other mechanical and biological factors might be effective in the osseointegration of the orthopedic implants?

**Q2.18** What other feasible reasons could be discussed for response of the cortical and trabecular responses to the implanted artificial device at the peri-implant and distant bone tissues.

## REFERENCES

[1] B. Peter, et al., Calcium phosphate drug delivery system: influence of local zoledronate release on bone implant osteointegration, Bone 36 (1) (2005) 52–60.
[2] D.R. Lionberger, P.C. Noble, Celecoxib does not affect osteointegration of cementless total hip stems, J. Arthroplast. 20 (2005) 115–122.
[3] Y.S. Park, et al., Correlation between microtomography and histomorphometry for assessment of implant osseointegration, Clin. Oral Implants Res. 16 (2) (2005) 156–160.
[4] M. Monjo, et al., Correlation between molecular signals and bone bonding to titanium implants, Clin. Oral Implants Res. 24 (9) (2013) 1035–1043.
[5] J.L.C. Guirado, et al., Different configuration of socket shield technique in peri-implant bone preservation: an experimental study in dog mandible, Ann. Anat. 208 (2016) 109–115.
[6] S.V. Kellesarian, et al., Does local ibandronate and/or pamidronate delivery enhance osseointegration? A systematic review, J. Prosthodont. 27 (3) (2016) 240–249.
[7] D. Botticelli, N.P. Lang, Dynamics of osseointegration in various human and animal models-a comparative analysis, Clin. Oral Implants Res. 28 (6) (2017) 742–748.
[8] G.N. Thalji, S. Nares, L.F. Cooper, Early molecular assessment of osseointegration in humans, Clin. Oral Implants Res. 25 (11) (2014) 1273–1285.
[9] M. Cassetta, et al., Early peri-implant bone loss: a prospective cohort study, Int. J. Oral Maxillofac. Surg. 44 (9) (2015) 1138–1145.
[10] B.G. Beutel, et al., Implant design and its effects on osseointegration over time within cortical and trabecular bone, J. Biomed. Mater. Res. B Appl. Biomater. 104 (6) (2016) 1091–1097.
[11] P.H. Morberg, et al., Improved long-term bone-implant integration Experiments in transgenic mice overexpressing bovine growth hormone, Acta Orthop. Scand. 68 (4) (1997) 344–348.
[12] Z. Li, et al., In vivo monitoring of bone architecture and remodeling after implant insertion: the different responses of cortical and trabecular bone, Bone 81 (2015) 468–477.
[13] F. Morelli, et al., Influence of bone marrow on osseointegration in long bones: an experimental study in sheep, Clin. Oral Implants Res. 26 (3) (2015) 300–306.
[14] R. Nedir, et al., Predicting osseointegration by means of implant primary stability, Clin. Oral Implants Res. 15 (5) (2004) 520–528.
[15] A.E. Al-Subaie, et al., Propranolol enhances bone healing and implant osseointegration in rats tibiae, J. Clin. Periodontol. 43 (12) (2016) 1160–1170.
[16] L.G. Persson, et al., Re-osseointegration after treatment of peri-implantitis at different implant surfaces, Clin. Oral Implants Res. 12 (6) (2001) 595–603.

[17] S. Galli, et al., The effect of osteotomy dimension on osseointegration to resorbable media-treated implants: a study in the sheep, J. Biomater. Appl. 29 (8) (2015) 1068–1074.
[18] M.A. Alfarsi, S.M. Hamlet, S. Ivanovski, The effect of platelet proteins released in response to titanium implant surfaces on macrophage pro-inflammatory cytokine gene expression, Clin. Implant. Dent. Relat. Res. 17 (6) (2015) 1036–1047.
[19] T. Nagel, D.J. Kelly, Computational mechanobiology in cartilage and bone tissue engineering: From cell phenotype to tissue structure, in: Computational Modeling in Tissue Engineering, Springer, 2012, pp. 341–377.
[20] A. Levchuk, R. Müller, In vivo validation of predictive models for bone remodeling and mechanobiology, in: Computer Models in Biomechanics, Springer, 2013, pp. 383–394.
[21] R. Ogawa, D.P. Orgill, Mechanobiology of cutaneous wound healing and scarring, in: Bioengineering Research of Chronic Wounds, Springer Berlin Heidelberg, 2009, pp. 31–42.
[22] J. Gao, J.L. Williams, E. Roan, Multiscale modeling of growth plate cartilage mechanobiology, Biomech. Model. Mechanobiol. 16 (2) (2017) 667–679.
[23] S. Checa, D.P. Byrne, P.J. Prendergast, Predictive modelling in mechanobiology: Combining algorithms for cell activities in response to physical stimuli using a lattice-modelling approach, in: Computer Methods in Mechanics, Springer, 2010, pp. 423–435.
[24] A. Schwartz, S. Thomopoulos, The role of mechanobiology in the attachment of tendon to bone, in: Structural Interfaces and Attachments in Biology, Springer, 2013, pp. 229–257.
[25] S. Thomopoulos, The role of mechanobiology in the attachment of tendon to bone, IBMS BoneKey 8 (6) (2011) 271–285.
[26] A. Vetter, et al., The spatio-temporal arrangement of different tissues during bone healing as a result of simple mechanobiological rules, Biomech. Model. Mechanobiol. 11 (1) (2012) 147–160.
[27] G. Chen, M. Schuetz, M. Pearcy, Mechanobiology of bone development and computational simulations, in: Bone and Development, Springer, 2010, pp. 279–295.
[28] F. Vermolen, et al., A suite of mathematical models for bone ingrowth, bone fracture healing and intra-osseous wound healing, in: Advanced Computational Methods in Science and Engineering, Springer, 2009, pp. 289–314.
[29] A. Idhammad, A. Abdali, N. Alaa, Computational simulation of the bone remodeling using the finite element method: an elastic-damage theory for small displacements, Theor. Biol. Med. Model. 10 (1) (2013) 32.
[30] N.M. Harrison, et al., Failure modelling of trabecular bone using a non-linear combined damage and fracture voxel finite element approach, Biomech. Model. Mechanobiol. 12 (2) (2013) 225–241.
[31] A.D. Speirs, et al., Physiologically based boundary conditions in finite element modelling, J. Biomech. 40 (10) (2007) 2318–2323.
[32] M.M. Saunders, Small-scale mechanical testing: Applications to bone biomechanics and mechanobiology, in: Time Dependent Constitutive Behavior and Fracture/Failure Processes, 3 2011, pp. 345–352.
[33] A.I. Birkhold, et al., Mineralizing surface is the main target of mechanical stimulation independent of age: 3D dynamic in vivo morphometry, Bone 66 (2014) 15–25.
[34] A.I. Birkhold, et al., Monitoring in vivo (re) modeling: a computational approach using 4D microCT data to quantify bone surface movements, Bone 75 (2015) 210–221.
[35] B. Busse, et al., Increased calcium content and inhomogeneity of mineralization render bone toughness in osteoporosis: mineralization, morphology and biomechanics of human single trabeculae, Bone 45 (6) (2009) 1034–1043.

[36] E. Degenkolbe, et al., Improved bone defect healing by a superagonistic GDF5 variant derived from a patient with multiple synostoses syndrome, Bone 73 (2015) 111–119.
[37] N. Donos, X. Dereka, N. Mardas, Experimental models for guided bone regeneration in healthy and medically compromised conditions, Periodontol. **68** (1) (2015) 99–121.
[38] B. Faensen, et al., Aprotinin application has no negative effect on osseous implant integration: a biomechanical and histomorphometric investigation in a rat model, Arch. Orthop. Trauma Surg. 129 (1) (2009) 51–56.
[39] M. Fini, et al., Osteoporosis and biomaterial osteointegration, Biomed. Pharmacother. 58 (9) (2004) 487–493.
[40] R.M. Hoerth, et al., Mechanical and structural properties of bone in non-critical and critical healing in rat, Acta Biomater. 10 (9) (2014) 4009–4019.
[41] T.M. Keaveny, W.C. Hayes, A 20-year perspective on the mechanical properties of trabecular bone, J. Biomech. Eng. 115 (4B) (1993) 534–542.
[42] M. Kerschnitzki, et al., The organization of the osteocyte network mirrors the extracellular matrix orientation in bone, J. Struct. Biol. 173 (2) (2011) 303–311.
[43] C. Khodadadyan-Klostermann, et al., Distribution ofbone mineral density with age and gender in the proximal tibia, Clin. Biomech. 19 (4) (2004) 370–376.
[44] I. Könnecke, et al., T and B cells participate in bone repair by infiltrating the fracture callus in a two-wave fashion, Bone 64 (2014) 155–165.
[45] M. Libicher, et al., Osseous integration of calcium phosphate in osteoporotic vertebral fractures after kyphoplasty: initial results from a clinical and experimental pilot study, Osteoporos. Int. 17 (8) (2006) 1208–1215.
[46] M. Mullender, et al., Mechanotransduction of bone cellsin vitro: mechanobiology of bone tissue, Med. Biol. Eng. Comput. 42 (1) (2004) 14–21.
[47] H. Razi, et al., Skeletal maturity leads to a reduction in the strain magnitudes induced within the bone: a murine tibia study, Acta Biomater. 13 (2015) 301–310.
[48] A. Sharenkov, et al., Automatic initial contact detection during overground walking for clinical use, Gait Posture 40 (4) (2014) 730–734.
[49] P. Strube, et al., Influence of age and mechanical stability on bone defect healing: age reverses mechanical effects, Bone 42 (4) (2008) 758–764.
[50] B. Trepczik, et al., Endochondral ossification in vitro is influenced by mechanical bending, Bone 40 (3) (2007) 597–603.
[51] B.M. Willie, et al., Diminished response to in vivo mechanical loading in trabecular and not cortical bone in adulthood of female C57Bl/6 mice coincides with a reduction in deformation to load, Bone 55 (2) (2013) 335–346.
[52] W. Whitehouse, E. Dyson, C. Jackson, A fine structure in the trabeculation of the human rib, Br. J. Radiol. 44 (521) (1971) 367–372.
[53] W. Whitehouse, Scanning electron micrographs of cancellous bone from the human sternum, J. Pathol. 116 (4) (1975) 213–224.
[54] W. Whitehouse, E. Dyson, C. Jackson, The scanning electron microscope in studies of trabecular bone from a human vertebral body, J. Anat. 108 (Pt 3) (1971) 481.
[55] H.K. Uhthoff, Z. Jaworski, Bone loss in response to long-term immobilisation, J. Bone Joint Surg. 60 (B(3)) (1978) 420–429.
[56] E. Bassey, et al., Pre-and postmenopausal women have different bone mineral density responses to the same high-impact exercise, J. Bone Miner. Res. 13 (12) (1998) 1805–1813.
[57] R.K. Fuchs, J.J. Bauer, C.M. Snow, Jumping improves hip and lumbar spine bone mass in prepubescent children: a randomized controlled trial, J. Bone Miner. Res. 16 (1) (2001) 148–156.
[58] A. Heinonen, et al., Randomised controlled trial of effect of high-impact exercise on selected risk factors for osteoporotic fractures, Lancet 348 (9038) (1996) 1343–1347.

[59] T. Kato, et al., Effect of low-repetition jump training on bone mineral density in young women, J. Appl. Physiol. 100 (3) (2006) 839–843.
[60] A. Vainionpää, et al., Effects of high-impact exercise on bone mineral density: a randomized controlled trial in premenopausal women, Osteoporos. Int. 16 (2) (2005) 191–197.
[61] A. Vainionpää, et al., Intensity of exercise is associated with bone density change in premenopausal women, Osteoporos. Int. 17 (3) (2006) 455–463.
[62] D. Kerr, et al., Exercise effects on bone mass in postmenopausal women are site-specific and load-dependent, J. Bone Miner. Res. 11 (2) (1996) 218–225.
[63] W.M. Kohrt, A.A. Ehsani, S.J. Birge, Effects of exercise involving predominantly either joint-reaction or ground-reaction forces on bone mineral density in older women, J. Bone Miner. Res. 12 (8) (1997) 1253–1261.
[64] W.M. Kohrt, et al., Additive effects of weight-bearing exercise and estrogen on bone mineral density in older women, J. Bone Miner. Res. 10 (9) (1995) 1303–1311.
[65] E. Seeman, Modeling and remodeling: the cellular machinery responsible for the gain and loss of bone's material and structural strength, Princ. Bone Biol. 1 (2008) 3–28.
[66] V. Glatt, et al., Age-related changes in trabecular architecture differ in female and male C57BL/6J mice, J. Bone Miner. Res. 22 (8) (2007) 1197–1207.
[67] L.J. Melton, et al., Does reduced skeletal loading account for age-related bone loss? J. Bone Miner. Res. 21 (12) (2006) 1847–1855.
[68] J. Somerville, et al., Growth of C57BL/6 mice and the material and mechanical properties of cortical bone from the tibia, Calcif. Tissue Int. 74 (5) (2004) 469–475.
[69] K. Soballe, et al., Gap healing enhanced by hydroxyapatite coating in dogs, Clin. Orthop. Relat. Res. 272 (1991) 300–307.
[70] T. Moilanen, et al., Hydroxyapatite coating of an acetabular prosthesis, J. Bone Joint Surg. (Br.) 78 (2) (1996) 200–205.
[71] I. Onsten, et al., Migration of acetabular components, inserted with and without cement, in one-stage bilateral hip arthroplasty. a controlled, randomized study using roentgenstereophotogrammetric analysis, JBJS 76 (2) (1994) 185–194.
[72] Y. Gao, et al., The effect of surface immobilized bisphosphonates on the fixation of hydroxyapatite-coated titanium implants in ovariectomized rats, Biomaterials 30 (9) (2009) 1790–1796.
[73] T. Andersson, et al., Surface immobilized zoledronate improves screw fixation in rat bone: a new method for the coating of metal implants, J. Mater. Sci. Mater. Med. 21 (11) (2010) 3029–3037.
[74] J. Abtahi, P. Tengvall, P. Aspenberg, Bisphosphonate coating might improve fixation of dental implants in the maxilla: a pilot study, Int. J. Oral Maxillofac. Surg. 39 (7) (2010) 673–677.
[75] F. Agholme, et al., Local bisphosphonate release versus hydroxyapatite coating for stainless steel screw fixation in rat tibiae, J. Mater. Sci. Mater. Med. 23 (3) (2012) 743–752.
[76] S.-J. Lee, et al., Effect of bisphosphonates on anodized and heat-treated titanium surfaces: an animal experimental study, J. Periodontol. 82 (7) (2011) 1035–1042.
[77] M. Nepal, et al., Evaluation of osseointegration around tibial implants in rats by ibandronate-treated nanotubular ti-32nb-5zr alloy, Biomol. Ther. 22 (6) (2014) 563.
[78] H.J. Rønold, J.E. Ellingsen, The use of a coin shaped implant for direct in situ measurement of attachment strength for osseointegrating biomaterial surfaces, Biomaterials 23 (10) (2002) 2201–2209.
[79] W. Becker, et al., Histologic evaluation of implants following flapless and flapped surgery: a study in canines, J. Periodontol. 77 (10) (2006) 1717–1722.
[80] A. Piattelli, A. Scarano, M. Piattelli, Detection of alkaline and acid phosphatases around titanium implants: a light microscopical and histochemical study in rabbits, Biomaterials 16 (17) (1995) 1333–1338.

[81] M. Monjo, et al., In vivo expression of osteogenic markers and bone mineral density at the surface of fluoride-modified titanium implants, Biomaterials 29 (28) (2008) 3771–3780.
[82] P.V. Hauschka, et al., Osteocalcin and matrix Gla protein: vitamin K-dependent proteins in bone, Physiol. Rev. 69 (3) (1989) 990–1047.
[83] C. Desbois, D.A. Hogue, G. Karsenty, The mouse osteocalcin gene cluster contains three genes with two separate spatial and temporal patterns of expression, J. Biol. Chem. 269 (2) (1994) 1183–1190.
[84] P. Ducy, et al., Increased bone formation in osteocalcin-deficient mice, Nature 382 (1996) 448–452.
[85] T. Berglundh, et al., De novo alveolar bone formation adjacent to endosseous implants, Clin. Oral Implants Res. 14 (3) (2003) 251–262.
[86] F. Rossi, et al., Bone-healing pattern at the surface of titanium implants: an experimental study in the dog, Clin. Oral Implants Res. 25 (1) (2014) 124–131.
[87] A. Quaranta, G. Iezzi, A. Scarano, P.G. Coelho, I. Vozza, M. Marincola, A. Piattelli, A histomorphometric study of nanothickness and plasma-sprayed calcium-phosphorous-coated implant surfaces in rabbit bone, J. Periodontol. 80 (2010) 556–561.

## FURTHER READING

[88] C.T. Rubin, L. Lanyon, Regulation of bone formation by applied dynamic loads, J. Bone Joint Surg. 66 (3) (1984) 397–402.
[89] P.-C. Chang, N.P. Lang, W.V. Giannobile, Evaluation of functional dynamics during osseointegration and regeneration associated with oral implants, Clin. Oral Implants Res. 21 (2010) 1–12.

## CHAPTER 3

# Osseointegration Concepts of Trauma Fracture Fixation and Joint Replacement

## 3.1 INTRODUCTION

Fixation of bone fractures is managed by the utilization of trauma implants. The created bone fragments are positioned in anatomical alignment to facilitate the formation of bone tissue at the fracture gaps, from which the bone fracture is healed. Bone healing of trauma fracture fixation is a mechanical-biological process at which the implant would be adequately interlocked with the bone to provide biological union of bone fragments at the fracture site. In view of osseointegration concept, biological integration of trauma screws is greatly beneficial to enhance the mechanical interlocking of the screw to the bone. In fact, early integration of the bone and screws would promote stability of the fracture fixation during biological healing of the fracture. Strong purchasing of the screw inside the bone would cause generation of bending stress through the implant. If this bending stress would not be effectively transferred to the implant, poor shear strength between the bone and screw would induce dislocation and ultimately loosening of the implant. It is worth mentioning that fixation of trauma implants (e.g., such as plate, intramedullary nail, or external fixator) to the bone is managed by screws in various types and sizes; therefore, integration of screws to the bone is the key factor of trauma implant stability for the treatment of bone fractures.

Joint replacement is the other type of implant, which is permanently replaced to the body for the treatment of joint severe injuries. Body movement is relied on the joints between the bones. Existence of cartilage (a connective tissue to the bone tissue at the joint which is softer than the bone tissue) at the joints would provide moving of the bone through each other while constrained by muscles, tendons, and ligaments. Due to the excessive loading conditions at the joints (e.g., increased body weight) and low level of daily exercises, the cartilage tissue might be affected. Increased contact stress

*Osseointegration of Orthopaedic Implants*
https://doi.org/10.1016/B978-0-12-813384-2.00003-5

at the joint may cause degeneration of the cartilage, which results in the resorption of the cartilage in contact areas with higher amount of contact stress. Lack of cartilage tissue is associated with contacting of the bone-to-bone tissue, which cause severe pain by the movement of the bones through each other at the joint. Cartilage tissue would not be regenerated and therefore, the replacement of prosthesis at the joint has been extensively used for the treatment of osteoarthritis. The joint implant (which is also known as joint prosthesis) consists of several components which are positioned at the connective bones at the joint while provides physiological range of motion (ROM).

In this chapter, these two types of orthopedic implants are reviewed in view of osseointegration aspects. In first part of the chapter, trauma implants are expressed generally and then the various aspects of the screw integration to the bone tissue are discussed. It is attempted to highlight the mechanical and biological challenges of the screw insertion through the bone. In the second part of the chapter, joint replacements are reviewed in general aspects such as material, biomechanical evaluation methods, design features, and types of implantation to the bone. Through which, the osseointegration of the joint replacements is discussed. Long-term integration of the joint replacement prostheses to the bone would be currently a major issue to increase the life of joint replacement at the joints. Other types of implants (spinal implants and bone tissue scaffolds) are reviewed in Chapter 4.

## 3.2 OSSEOINTEGRATION OF TRAUMA FRACTURE FIXATION

### 3.2.1 General Aspects of Trauma Fracture Fixation

Trauma implants are normally manufactured from titanium alloy or stainless steel. These implants were first developed from stainless steel material. Today, various types of trauma implants (e.g., plating systems, intramedullary nails, and external fixations) have been developed for fracture fixation of all human bones from mandibular to foot bones. If the fracture is occurred in the middle diaphyseal portion of the long bones, all plating systems, intramedullary nails, and external fixations could be utilized. According to the severity and type of the fracture, each of these groups of trauma implants has specific advantages for fixation of fractures at the middle diaphyseal of long bones. In the case of fracture without fragmentary (cracking), external fixators along with adequate immobilizing of the fracture site would be usable for healing of created crack to the bone. For such fracture conditions, using of external fixator prevents open operation at the fracture site. In occurrences of fragmentary fractures in the middle diaphyseal of long bones with high-load bearing conditions (high-load bearing is the presence of high amount of bending, compression, or torsional stresses to the bone), such as femur, tibia, humerus bones, and intramedullary nail could have better biomechanical advantages compared to external fixators. In this type of fracture condition in long bones with medium load bearing condition or in bones with lower width of the cortical bone than high-load bearing bones (e.g., radius, ulna, clavicle, metacarpal, metatarsal, etc.), plating systems have biomechanical, biological, and clinical advantages than intramedullary nailing and external fixation.

The fracture might be fragmented at the metaphyseal portion of the long bones (metaphyseal bone is the portion of the long bone near to the joints at which the majority of the bone volume is trabecular or cancellous bone). Fracture fixation of such fracture is managed by plating systems or cannulated screws. This is whereas, utilization of intramedullary nail or external fixators is not recommended and in majority of cases would not be applicable. Normally, multiple screws are essential to be inserted through the created fragments to restore them in anatomical alignment. Two or multiple types of trauma implants are utilized in some cases. For instance, in open fractures [open fracture is when the bone fragment is severely displaced and damage soft tissues or even cause soft tissue (e.g., tendon, ligament, and skin) rupture], the damaged or rupture soft tissue is first treated to reduce the non-healing risk of ruptured soft tissues which might be arisen due to the delay in the treatment. In order to prevent further bleeding and soft tissue damage caused by created bone fragments, external fixator is utilized for temporary fixation of fracture during the soft tissue treatment. On completion of soft tissue treatment, which may take couple of weeks or months, plating or intramedullary nailing fixation methods are utilized for effective fixation and healing of the fracture and external fixator might be removed if extra stability of the fracture fixation would not be required. In condition of open fracture with long fracture length, the external fixator might be kept along with the plating fixation to enhance the rigidity of the fracture fixation for better biomechanical and clinical outcomes.

### 3.2.2 Osseointegration of Trauma Screw

Screw attachment or integration to the bone has been validated using resonance frequency measurement (RFM), removable torque test (RTV), biomechanical testing, and ultrastructural examination methods. Normally, the screw is implanted into the animal bones and the evaluation record is carried out in regular time intervals after implantation [1–4]. As highlighted by Buchter et al. [1], adequate osseointegrated screw would enhance its stability during postimplantation period which might be influenced by surgical procedure, implant-bone interface conditions, and morphology of the implant in micro and macroscales. Investigation of screw integration to the bone is valuable and significantly helps for better understanding of screw stability during implantation period. However, such researches have been performed for the evaluation of dental implants. Although these outcomes are beneficial and applicable for trauma screws, organization of specific osseointegration investigations for the implantation of trauma implants into the animal bones would be crucial for effective studying of the bone attachment conditions with trauma screws. For instance, in trauma plating fixation, compression stress is transferred to the screws from the stabilized bone fragments, which induces the generation of bending stress on the plate. By increasing of the compression loading, generated bending stress on the plate would tend to move out the screws from the bone, from which shear stress at the bone-screw interface is sharply increased. Therefore, considering of this loading mechanism in the evaluation of trauma screws could better describe the osseointegration conditions after the implantation of trauma implants.

The presence of the threads on the surface of the screw would allow gripping or tapping of the screw while purchasing or inserting to the bone. The compact cortical bone would be effectively gripped through the screw threads, while the spongy trabecular bone might not be perfectly penetrated to the screw threads. Thus, the longer pitch and deeper depth of the threads are desired for the enhancement of the contact surface between the trabecular bone and screw, while shorter pitch and lesser depth of the threads are desired for increase of the tapping strength between the cortical bone and screw. On the other hand, sharper thread head and higher helix angle are advantageous for easier purchasing or insertion of the screw through the bone. Trauma screws might be designed single, double, triple, or even quad threads. Increasing the number of tread lines in the pitch of the first thread line would allow higher helix angle with lower pitch (in normal single thread line, increase of the helix angle would increase the thread pitch which might not be desired due to the reduction of the shear strength between the bone and screw). Overall, the geometry of the screw threads would have a significant effect on the mechanical strength of the contact between the bone and screw. Likewise, the adequate compatibility of the screw threads with

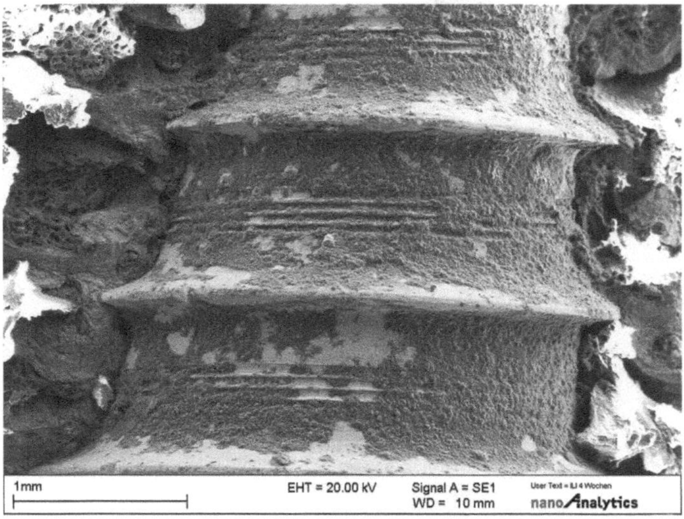


**Fig. 3.1** Integration of the bone and screw, adhesion, and the formation of the bone could be observed at the contact surfaces of the screw threads with the bone. *(With courtesy of A. Büchter, et al., Biological and biomechanical evaluation of interface reaction at conical screw-type implants. Head Face Med. 2 (2006) 5, and BioMed Central.)*

the morphology of the bone would enhance the load distribution or level of the loaded strain to the peri-implant bone tissue from which the physiological mechanical stimulation signals for effective bone remodeling process is established at the bone-implant interface. This would increase the formation of mineralized new woven bone tissue at the contact surfaces between the bone and screw at the depth of the threads. Fig. 3.1 displays the integration of the screw and bone at which the bone tissue would be formed through the depth of the threads. Morphology of the screw profile at the threaded portion was reported to have a significant effect in the early osseointegration [1, 2, 5]. Parabolic shape of screw (Fig. 3.2) could enhance micromotion in the range of 500–3000 μm through the bone tissue at the interface of bone-implant [6–9], thus better mechanical stimulation of the osteogenic cells to be formed as the lamellar bone. Osseointegration aspects of the trauma screws are summarized in Table 3.1.

### 3.2.3 Osseointegration of Pedicle Screw

In some cases of screw fixation, the threaded length of the screw shaft has been found significant in effective bone-screw integration. Loosening and breakage of the pedicle screw in transpedicular stabilization fixation of spinal

**Fig. 3.2** Parabolic shape of the screw, this type of screw threads has been found effective in better distribution or transferring of the localized strain level at the interface of the bone and screw from which the mechanical stimulation for mineralization of the osteocyte cells and ultimately formation of the woven and lamellar bone would be enhanced. *(With courtesy of A. Büchter, et al., Biological and biomechanical evaluation of interface reaction at conical screw-type implants. Head Face Med. 2 (2006) 5, and BioMed Central.)*

**Table 3.1** Osseointegration aspects of the screw

| **Aspect** | **Description** |
| --- | --- |
| Evaluation methods | Methods such as resonance frequency measurement (RFM), removable torque test (RTV), biomechanical testing, and ultra-structural examination are utilized to evaluate the extent of integration between the bone and the screw. The methods for evaluation of implant osseointegration have been reviewed with more details for various types of orthopedic implant in Chapter 2. |

**Table 3.1** Osseointegration aspects of the screw—cont'd

| Aspect | Description |
|---|---|
| General effective parameters on osseointegration | Generally, the parameters such as surgical procedure, implant-bone interface conditions, and morphology of the implant in micro and macroscales are greatly influenced on integration of the screw to the bone, particularly the trabecular bone. |
| Shear stress at the bone-screw interface | Higher shear strength at the interface of the bone and the implant would be contemplated as the biomechanical parameter in the enhancement of the screw design and surface conditions. |
| Gripping or tapping of the screw threads to the bone | One of the important factors in placement of the screw to the bone is the self-tapping ability of the screw while twisting of the screwdriver. It means that with creation of the prehole in smaller diameter, the screw tip would create the thread line while inserting to the bone without using of the tapper tool before screw insertion. In the case of high level of compactness of cortical bone or thick cortical layer, it is advantageous to enhance the self-drilling ability of the screw to have sharp flutes at the tip. Tapping of the screw would be completed with adequate gripping of the screw at the cortical layer after complete insertion of the screw. |
| Pitch and depth of the threads | Longer pitch and deeper depth of the threads are desired for enhancement of the contact surface between the trabecular bone and the screw, while shorter pitch and lesser depth of the threads are desired for increase of the tapping strength between the cortical bone and the screw. |
| Crest and helix angle of the thread | Sharper thread crest and higher helix angle are advantageous for easier purchasing or insertion of the screw through the bone. |
| Number of thread lines | Number of tread lines in the pitch of the first thread line would allow higher helix angle with lower pitch (in normal single thread line, increase of the helix angle would increase the thread pitch which might not |

*Continued*

**Table 3.1** Osseointegration aspects of the screw—cont'd

| Aspect | Description |
|---|---|
| | be desired due to reduction of the shear strength between the bone and the screw. Trauma screws might be designed with single, double, triple, or even quad thread lines. |
| Compatibility of the screw threads with the bone morphology | Adequate compatibility of the screw threads with the morphology of the bone would enhance the load distribution or level of the loaded strain to the peri-implant bone tissue from which the physiological mechanical stimulation signals for effective bone remodeling process is established at the bone-implant interface. This would increase the formation of the mineralized new woven bone tissue at the contact surfaces between the bone and the screw at the depth of the threads. |
| Parabolic shape of screw | Parabolic shape of the screw (Fig. 3.2) could enhance the micromotion in the range of 500–3000 μm through the bone tissue at the interface of the bone-implant, which would be beneficial for better mechanical induction of the osteogenic cells to be formed as lamellar bone. |

surgeries has been widely reported [10–16]. This would be because of various physiological conditions that is generated during flexion-extension, lateral bending, and internal-external movement of the spinal column [17]. It has been achieved that the key factor for the reduction of pedicle screw loosening and breakage is the enhancement of screw osseointegration to the bone segments of spinal column. It is essential to promote biological and mechanical integration of screw-bone at the interface and enhance the fatigue strength of the screw neck [18]. Several methods such as bioactive coating of screw threads, augmentation of screw threads with the bone-cement, modification of screw threads, and expanding the screw after insertion have been developed [19–23] for improvement of the bone-implant integration at the vertebral body. According to the Tsuang's et al. [16] results, the pedicle screw with adequate threaded and unthreaded length could enhance the bone-implant integration.

Threaded portion of the pedicle screw in trabecular bone of the vertebrae provides a mechanical interlocking or trapping of the screw into the bone. Unthreaded portion of the pedicle screw shaft at the proximal zone could squeeze the bone chips and increase the friction between the screw and the cortical bone (bone chips are created during purchasing of the threaded portion into the bone). In fact, unthreaded length is designed based on the length of pedicle section of the vertebrae at which the width of trabecular bone is much lower than that of trabecular bone at body section of the vertebral bone (Fig. 3.3). Based on the performed biomechanical test by Tsuang et al. [16], adequate lengths of threaded and unthreaded could increase screw removal torque. It was achieved that one-third unthreaded length would have better integration of the pedicle screw and vertebrae than full threaded, half threaded, and one-third threaded length of the screw. The advantage of the threaded part of the pedicle screw is to squeeze the bone

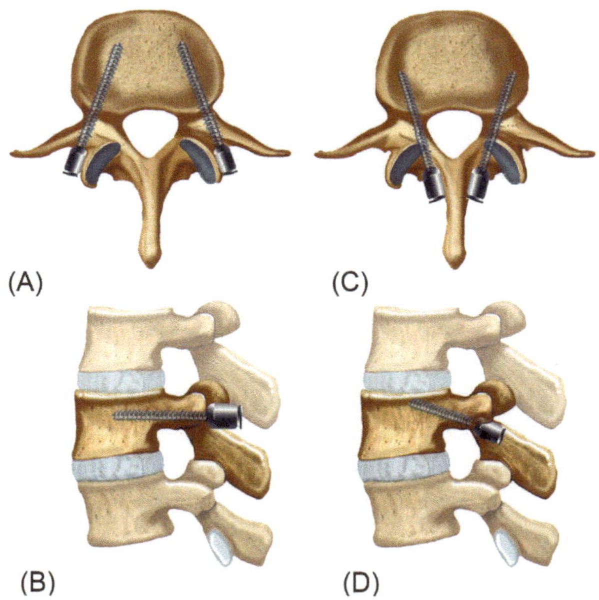


**Fig. 3.3** Schematic illustration of the pedicle screw through vertebral segments of the spinal column, tapping of the proximal portion of the screw at the pedicle isthmus has been found significantly crucial for enhanced stability of the bone-screw construct in the long-term implantation. This enhanced stability would deduct the risk of loosening or dislocation due to the presence of dynamic or cyclic physiological loading conditions during healing period. *(With courtesy of Y.-R. Chen, et al., Minimally invasive lumbar pedicle screw fixation using cortical bone trajectory—a prospective cohort study on postoperative pain outcomes, Cureus 8 (2016) e714.)*

chips and increase the bone-implant friction. The advantage of the friction effect is to enhance the fatigue strength of the pedicle screw. It was observed that the transferred physiological stress to the pedicle screw is the highest at the proximal portion, which is located at the pedicle section of the vertebrae. Unthreading of the proximal portion of the screw would significantly reduce the stress concentration factor and therefore, the fatigue strength of the screw is increased. It has been highlighted that around 60% of the shear strength between the pedicle screw and the vertebrae is supplied at the interface of the pedicle screw with pedicle section of the vertebrae. This is while 10%–15% of shear bearing is obtained at the body section and 20%–25% by purchasing of the screw through the far cortex [24].

The cortical bone at the pedicle section of the vertebrae has been found to be less compact compared to that of long bones. Therefore, the press-fit effect of unthreaded portion of the pedicle screw may increase the risk of crack creation to the pedicle isthmus during or even after screw implantation. In further details, when the pedicle screw is purchasing through the vertebrae, the crushed trabecular bone by the thread flute at the tip of the screw are gradually shifting back to the pedicle isthmus by complete screw insertion in place. These trabecular bone chips would increase the press-fit force of unthreaded part of the screw which may not be desired [25]. This might be the reason that the current commercial pedicle screws are developed without unthreaded part in the shaft. In order to provide the advantage of friction advantage of the screw at the pedicle isthmus, the proximal portion of the screw is threaded but with larger core diameter (or lower thread depth) and higher thread crest at the cortical section compared to trabecular section. This would allow engagement of the pedicle screw with the cortical bone at the pedicle isthmus while enhancing the friction of screw-bone with reducing the increased press-fit force at the pedicle isthmus and the risk of bone fracture at the pedicle isthmus.

## 3.3 OSSEOINTEGRATION OF JOINT REPLACEMENTS

### 3.3.1 General Aspects of Joint Replacements

Joint replacement consists of various prostheses or components. Of prostheses, two prostheses are articulated to provide natural ROMs at the affected joint. Other prostheses are implanted to the bones to establish integration of the joint replacement to the bone and its stability in the long-term implantation. Other supplementary prosthesis might be used to link the articulated and implanted prostheses. Total joint replacements have been developed from several materials. Cobalt chromium (CoCr) alloy [26–31] and ultrahigh-molecular-weight polyethylene (UHMWPE) are the key materials in the development of total joint replacements. Articulation of these materials with the consideration of specific material improvements processes could provide smooth movement with no contact wear. Although other articulation such as zirconium oxide/polyetheretherketone ($ZrO_2$-PEEK) has been introduced to be used in total joint replacements, articulation of CoCr-UHMWPE is still utilized in total joint replacements such as knee, hip, spinal disc, shoulder, and ankle replacements. Other than CoCr and UHMWPE, stainless steel and titanium alloy are used in the development of other components of joint replacements, which are inserted to the bone. In aspect of biomechanical evaluation, various dynamic and static requirements are essential for an effective validation of the joint replacements. Overall, mechanical strength of the joint prosthesis need to be evaluated under various types of loading conditions at the joint (e.g., compression, shear, torsion, and bending). Likewise, the contact stress and sliding distance are computationally or experimentally analyzed to predict the level of wear rate between the articulated prosthesis in the long-term use. Furthermore, the movement of articulated prosthesis through each other is examined in respect of constraints at the interface. The ROM in lateral-medial, anterior-posterior, flexion-extension, internal-external, inferior-posterior, and adduction-abduction axes is computationally and experimentally evaluated to be similar to natural ROMs in healthy joint. The design profile of articulated prostheses provides an essential range of force/displacement and torque/rotation during movement of the articulation components of the joint replacement.

### 3.3.2 Design Aspects of Joint Replacements

In view of the prosthesis design, the components of the joint replacement are modeled to be adopted with the anatomy of the bones at the joint [32–39]. The shape of the prosthesis component at the articulation interface is designed to provide natural physiological movement of the joint while dedicating the cartilage profile. On the other hand, the shape of the contact surfaces of the connective bones is modeled to be well matched with the surface profile of the articulation prostheses at the joint to provide effective transferring and distribution of the load between the articulated bones at the joint. Furthermore, the implanted prostheses to the bone are beneficial to prevent concentration of the physiological stress (through using of finite element analysis, physiological stress of the gait, stair climbing, etc. could be evaluated on the prosthesis) at the region of bone-prosthesis interface.

Implanted prostheses are enhanced with extended features to be interlocked with the bone [31, 40, 41]. The benefits of such features are to provide early mechanical integration of bone-prosthesis. This stability would induce effective mechanical stimulation of bone tissue in the long-term implantation. The design of extended features needs to be biomechanically optimized to allow expansive distribution of the load from the bone through the prosthesis. For instance, excessive stimulation of strain at the interface of bone-prosthesis has been addressed ($e > 50\,\mu m$) as one of the undesired factor that reduced osseointegration of tibial tray component (tibial tray is one of the component or prosthesis of total knee replacement) [42]. Fig. 3.4 would demonstrate the reviewed design aspects in glenoid component of the total shoulder replacement.

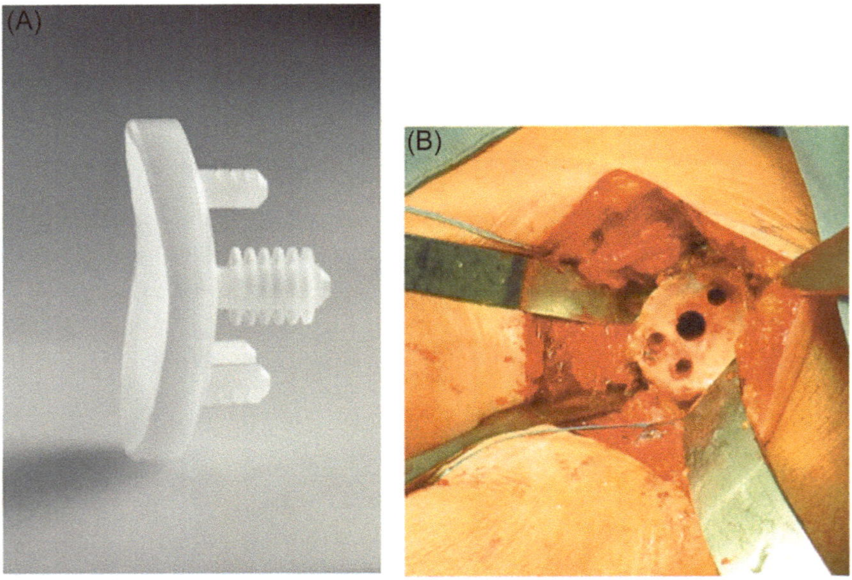


**Fig. 3.4** (A) Glenoid component of the total shoulder replacement and (B) preholes creation style for better placement of the prosthesis through the glenoid portion of the scapula bone. This would be one of the current design of the glenoid prosthesis with four extended features for the enhancement of the bone-prosthesis mechanical integration and optimal load transferring between the bone and the prosthesis from which the occurrences of the localized stress zones is significantly reduced compared to general design of this prosthesis with one or two extended tabs and normal profile. *(With courtesy of G. Merolla, et al. Total shoulder replacement using a bone ingrowth central peg polyethylene glenoid component: a prospective clinical and computed tomography study with short-to mid-term follow-up. Int. Orthop. 40 (11) (2016), 2355–2363, and getting permission from Springer Link.)*

### 3.3.3 Osseointegration of Joint Replacements

The placement of joint replacement prostheses to the bone is faced with various challenges of osseointegration in long-term replacement. Various concepts are studied for the replacement of joint prosthesis with effective and long-standing osseointegration. Osseointegration of the joint replacement could be assessed from different aspects of material, biomechanical, biological, and clinical requirements. Osseointegration of joint replacements is greatly correlated with the joint conditions after the replacement. Various factors would influence osseointegration of the joint replacement such as:

**(a).** inherent integration of the bone tissue to the prosthesis material at the bone-prosthesis interface,
**(b).** prosthesis general design,
**(c).** prosthesis features,
**(d).** bone quality,
**(e).** bone type at the bone-prostheses interface,
**(f).** contact conditions of articulated components,
**(g).** anatomic implantation of joint replacement components,
**(h).** mechanical strength of component under static and dynamic loading conditions,
**(i).** muscle-tendon strength, and
**(j).** ligament strength.

It could be emphasized that the assessment of joint conditions after replacement is complicated due to the involvement of many effective parameters for successful integration of the joint prosthesis to the bones at the affected joint.

### 3.3.4 Cemented Implantation of Joint Replacement Prosthesis

Currently, various concepts are considered for osseointegration enhancement of the joint replacement. Early mechanical interlocking or integration is the first requirement that facilitates biological integration of the bone through the prosthesis. As highlighted in Section 3.3.2, the shape of extended features is crucial to provide effective transferring or distributing of the load from the bone through the main body of the prosthesis. Using bone-cement for the replacement of the joint prosthesis would provide a connective layer between the bone and the prosthesis [43]. This is because the bone-cement could have better integration with the bone than the prosthesis. However, generated shear stress at the interface of the bone-cement could cause micro-cracking of the intermediate layer and debonding and loosening of the prosthesis in the long-term implantation (more than 5 years) in elderly patients and short-term implantation in youth patients (between 1 and 5 years implantation). Likewise, integration of the bone-cement and the prosthesis is brittle and might be affected by physiological dynamic loading conditions in the long-term implantation. To promote the bone-cement-prosthesis integration, surface roughening of the bone and the prosthesis was found to be effective to increase the shear strength at the interfaces of the bone-cement and cement-prosthesis. Due to porous structure of the bone tissue, particularly trabecular bone, cement could penetrate through the bone from which the intermediate layer at the interface of the bone-cement has higher shear strength than that of the prosthesis-cement interface. This fact was the key point of surface roughening of the implanted joint prosthesis to the bone. With application of coarse blasting, surface of metallic prosthesis was roughened to increase the thickness of the cement-prosthesis intermediate layer. However, penetration of the bone-cement through the prosthesis would not be adequately achieved as good as that of the cement-bone interface. Thereby, creating of thin porous layer on the surface of the prosthesis has been utilized to promote penetration and integration of the cement to the prosthesis.

### 3.3.5 Cementless Implantation of Joint Replacement Prosthesis

Although bone-cement is extensively used in the replacement of the joint prosthesis at the joint, however, cementless prostheses have been introduced for the enhancement of biological integration [44]. The main concept of cementless prosthesis is to establish the direct integration of the prosthesis to the bone without creating of the intermediate layer by the bone-cement. This concept attracted the orthopedic implant researchers and manufacturers to develop optional solutions. In order to achieve the direct biological integration at the bone-prosthesis interface, a thin bioactive layer on the surface of the prosthesis is needed to act as the catalyst or generator layer for the formation of intermediate layer at the interface. Indeed, this bioactive layer is hardly deposited to the surface of the prosthesis to establish the growth of the bone tissue to a thin layer of prosthesis at the bone-prosthesis interface. Porous coating of titanium and calcium phosphate-based materials has been found effective in the enhancement of the bone in-growth through the surface of the prosthesis. However, bonding or deposition strength of the coating layer has been found as a major challenge in the long-term replacement of the joint prosthesis to the bone. Therefore, various coating processes have been investigated for the establishment of a strong deposited and monotonic coating layer on the contact surfaces of the joint prosthesis with the bone. It could be highlighted that nonmonotonic distribution of the costing layer would cause its failure and the removal of coating layer. Effective deposition of the coating material through the prosthesis surface would be normally established with concentration or generation of the heat energy during the coating process, which causes melting of the thin layer of the prosthesis. However, the processing conditions would not be entirely under control and its effect may not be homogenous over the coating surface. The aim of coating deposition is to enhance in-growth of the bone tissue to the prosthesis in addition of the bone in-growth to the coating layer. The coating layer could be somehow considered as an intermediate layer between the bone and the prosthesis. This intermediate layer is first induced osteogenesis cells to generate new bone cells followed by mineralization of the osteocyte cells and the formation of the bone tissue on the surface of the prosthesis. Higher extent of bone apposition through the depth of the prosthesis surface would promote the osseointegration of the prosthesis. This process could be achieved effectively in patients with natural bone density and normal bone remodeling process.

### 3.3.6 Effect of Porous Layering on Osseointegration of Joint Replacements

Titanium and tantalum porous layering has been found with higher in-growth of the bone tissue at the interface of the bone-prosthesis [45]. More stable layer could be expected in the long-term replacement of prosthesis with titanium or tantalum porous surface compared to calcium-phosphate coating layer. Titanium is an osseointegrant material due to its high level of surface energy in nanoscale. Integration ability of the tantalum is even higher than titanium, which is currently used as the supplementary layer on the contact surfaces of the joint prostheses, which are in contact with the bone. Effectiveness of the bone in-growth in the long-term implantation is discussable from various aspects. Early formation of bone cellular matrix through the porous surface would be achieved due to the presence of a considerable level of strain stimulation ($50\,\mu m < \varepsilon < 100\,\mu m$) at the interface. However, effective bone remodeling process to resorb degenerated bone tissues and to regenerate new bone tissue inside the porous layer in the long-term implantation could be one of the challenges that might be expected. Likewise, regular or irregular geometrical structuring of the porous surface is other concern to establish the required distribution of the strain stimulation pattern.

Overall, osseointegration of the joint replacement is influenced by various factors, which would make its effectiveness with unexpected outcomes in the long-term implantation. This might be the reason that orthopedic surgeons recommend the life of less than 10 years for joint replacements treatment in patients with low level of activity.

## 3.4 SUMMARY

In this chapter, two types of orthopedic implants have been reviewed in general aspects. Various concepts of osseointegration that are associated with these implants are discussed accordingly. The first category was trauma implants which are used for fracture fixation of the bone in various parts of the body from skull to foot bones. Trauma implants are fixed to the bone fragments in anatomical position to allow healing of the bone fracture at the fracture gaps. Trauma implants are fixed to the bone by trauma screws. The integration of the screw to the bone is the osseointegration concern of this type of orthopedic implants. Therefore, the focus is to optimize the design and surface conditions of the screw threads and even the implantation methods for effective purchasing and tapping of the screw through the trabecular and cortical layers of the bone tissue. The design parameters such as pitch, depth, crest, helix angle, number of thread line, thread profile, general shaping of the thread, etc. are considered for the enhancement of the screw integration to the bone. On the other hand, biological integration of the screw with the bone tissue would be a crucial factor for the long-term stability of the screw through the bone tissue. In fact, the formation of the new lamellar bone at the interface of the bone and the screw is the ultimate target of the successful osseointegration of the trauma screws. Therefore, the design parameters are beneficial to be optimized for an effective mechanical stimulation of the osteogenic cells to be formed as a mature or lamellar bone at the bone-screw interface. Due to a similar concept of pedicle screw in the spinal stabilization system, its osseointegration challenges and concepts are reviewed in this chapter. Self-tapping of the pedicle screw at the initial step of the insertion and tapping of the screw through the cortical layer of the pedicle at the final step of the implantation are the main concerns of pedicle screw design. The design parameters of the pedicle screw are currently different at the distal and proximal portions (distal portion is compatible with the trabecular bone and proximal portion is compatible with the pedicle portion).

In the second part of this chapter, osseointegration of the joint replacements is reviewed. Generally, osseointegration of the joint replacements is in higher attention due to the high risk of implant loosening for the implantation of this type of orthopedic implants. Normally, the joint replacements are implanted by using of the bone-cement to establish an intermediate layer between the bone and the prosthesis. It is advantageous to create the porous structure layer on contact surfaces of the joint replacement prostheses with the bone. This would allow penetration of the pone cement through the prosthesis surface and thus prevent the debonding of the cement from the prosthesis due to the creation of the micro-cracks in the long-term implantation to the bone. Likewise, roughening of the bone would enhance the shear strength between the bone and the cement. One of the main difference between the joint prosthesis and trauma screws is that the trauma screws are entirely inserted through the bone from which the mechanical integration is effectively achieved. However, the joint replacement is press-fitted on the surface of the bones at the affected joint. Therefore, joint replacements are normally designed with several extended features to enhance the early and long-term mechanical integration of the prosthesis with the bone. Designing of this extended feature would be crucial for effective load transferring or distribution between the bone and the prosthesis, which has a significant effect on the prevention of implant loosening and reduction of contact stress at the articulation components of the joint replacements. Due to osseointegrant or osseoconductive ability of the titanium and tantalum materials, coating of these metallic materials on the surface of the cobalt-chromium and stainless steel components of the joint replacement in contact surfaces with the bone is advantageous for the enhancement of the bone in-growth through the surface of the cementless prosthesis.

Further concepts of osseointegration and the methods that would be utilized for the enhancement of osseointegration are reviewed and discussed in Chapters 5–7. Studying of these chapters is highly recommended for complete review of osseointegration in trauma and joint replacements implants.

## 3.5 REMIND AND LEARN

**Q3.1** What is the indication of the trauma implants? What types of trauma implants have been developed? How these implants would treat the indicated bone injuries?

**Q3.2** Express the mechanical purchasing of the trauma screw to the bone. How screw loosening might be occurred? (Consider the mechanical interaction between the trauma screw and trauma implant in your answer.)

**Q3.3** Express the fracture types based on the location of the bone fracture. What types of trauma implant might be used for fracture fixation of middle diaphyseal portion of the long bones?

**Q3.4** Explain the fragmentary and nonfragmentary fracture at the middle diaphyseal portion of the long bones. Describe what type of trauma implants could be used for each fracture type.

**Q3.5** How the fractures at the metaphyseal portion of the long bones are treated with trauma implants? Discuss why the intramedullary nail might not be adequate for this type of fracture.

**Q3.6** In what condition two or multiple trauma implants might be utilized for fixation of bone fracture? In what condition, external fixator might be kept along with the plating fixation in the treatment of bone fracture?

**Q3.7** What type of experiments or methods could be organized for evaluation the implant osseointegration?

**Q3.8** Describe why the concept of osseointegration evaluation need to be different with what have been experienced with dental implants.

**Q3.9** What design parameters of the screw threads are influencing on integration of the screw to the bone? Express the effect of these parameters in details.

**Q3.10** Explain why osseointegration of the pedicle screw would be crucial in spinal stabilization treatment. Which part of the pedicle screw could be improved to reduce the feasible failures of spinal stabilization treatment?

**Q3.11** What methods could be utilized for the enhancement of pedicle screw osseointegration with the vertebral bone?

**Q3.12** Discuss how the unthreaded portion of the pedicle screw at the proximal section could promote the osseointegration of this screw. What are the disadvantages of the unthreaded proximal portion of the pedicle screw? What is the best design of pedicle screw for a secure and effective purchasing through the vertebral bone?

**Q3.13** Express the general aspects of the joint replacements in view of included prostheses, materials, and biomechanical evaluation methods.

**Q3.14** What design parameters of prostheses are greatly crucial for a successful performance of the joint replacements? What is the benefit of extended features of implanted prostheses of the joint replacements?

**Q3.15** What factors could be found effective on osseointegration of the joint replacement?

**Q3.16** What are advantages and disadvantages of the bone-cement for the implantation of the implanted prosthesis of the joint replacements?

**Q3.17** What methods could be used to enhance the effectiveness of the cemented implantation of the joint replacements in view of osseointegration? Express your answer in details.

**Q3.18** Explain the concept of the cementless implantation of joint replacements. What solutions have been introduced for the development of cementless joint replacements?

**Q3.19** Discuss the effect of porous layering on osseointegration of joint replacements.

## 3.6 THINK AND DISCUSS

**Q3.20** Characteristics of the osseoconductive layer on the surface of the trauma screws and implanted prostheses of the joint replacements are technically different. Discuss these differences with adequate reasons.

**Q3.21** Discuss why the osseointegration of the joint replacements is greatly important compared to osseointegration of the trauma screws. Consider mechanical and biological integration aspects in your discussion.

**Q3.22** Osseointegration of trauma plating fixation could be developed by the new concept of Advance Healing Fixation System (AHealFS) which has been published in the book "Trauma Plating Fixation." Provide a research in the mentioned book and discuss how this new concept could enhance the osseointegration of the trauma screws.

## REFERENCES

[1] A. Büchter, et al., Biological and biomechanical evaluation of interface reaction at conical screw-type implants, Head Face Med 2 (1) (2006) 5.
[2] T. Albrektsson, et al., Osseointegrated titanium implants: requirements for ensuring a long-lasting, direct bone-to-implant anchorage in man, Acta Orthop. Scand. 52 (2) (1981) 155–170.
[3] T. Albrektsson, et al., The interface zone of inorganic implants in vivo: titanium implants in bone, Ann. Biomed. Eng. 11 (1) (1983) 1–27.
[4] T. Albrektsson, et al., Osseointegrated oral implants: a Swedish multicenter study of 8139 consecutively inserted Nobelpharma implants, J. Periodontol. 59 (5) (1988) 287–296.
[5] D. Buser, et al., Influence of surface characteristics on bone integration of titanium implants. A histomorphometric study in miniature pigs, J. Biomed. Mater. Res. A 25 (7) (1991) 889–902.
[6] U. Joos, D. Vollmer, J. Kleinheinz, Effect of implant geometry on strain distribution in peri-implant bone, Mund Kiefer Gesichtschir. 4 (3) (2000) 143–147.
[7] U. Meyer, T. Meyer, D. Jones, Attachment kinetics, proliferation rates and vinculin assembly of bovine osteoblasts cultured on different pre-coated artificial substrates, J. Mater. Sci. Mater. Med. 9 (6) (1998) 301–307.
[8] U. Meyer, et al., Experimentelle und Finite-Elemente-Analyse der Biomechanik des Unterkiefers unter Belastung, Mund Kiefer Gesichtschir. 4 (1) (2000) 14–20.
[9] U. Meyer, et al., Sensitivity analysis of bone geometries around oral implants upon bone loading using finite element method, Comput. Methods Biomech. Biomed. Engin. 3 (2001) 553–559.
[10] H. Pihlajamäki, P. Myllynen, O. Böstman, Complications of transpedicular lumbosacral fixation for non-traumatic disorders, J. Bone Joint Surg. (Br.) 79 (2) (1997) 183–189.
[11] S.H. Davne, D.L. Myers, Complications of lumbar spinal fusion with transpedicular instrumentation, Spine 17 (6) (1992) S184–S189.
[12] J.E. Lonstein, et al., Complications associated with pedicle screws, JBJS 81 (11) (1999) 1519–1528.
[13] K. Okuyama, et al., Can insertional torque predict screw loosening and related failures?: an in vivo study of pedicle screw fixation augmenting posterior lumbar interbody fusion, Spine 25 (7) (2000) 858–864.
[14] A.W. Kwok, et al., Insertional torque and pull-out strengths of conical and cylindrical pedicle screws in cadaveric bone, Spine 21 (21) (1996) 2429–2434.
[15] C.C. Hsu, et al., Increase of pullout strength of spinal pedicle screws with conical core: biomechanical tests and finite element analyses, J. Orthop. Res. 23 (4) (2005) 788–794.
[16] F.-Y. Tsuang, et al., Biomechanical arrangement of threaded and unthreaded portions providing holding power of transpedicular screw fixation, Clin. Biomech. 39 (2016) 71–76.
[17] C.-S. Chen, et al., Stress analysis of the disc adjacent to interbody fusion in lumbar spine, Med. Eng. Phys. 23 (7) (2001) 485–493.
[18] C.-K. Chao, et al., Increasing bending strength and pullout strength in conical pedicle screws: biomechanical tests and finite element analyses, Clin. Spine Surg. 21 (2) (2008) 130–138.
[19] M.H. Krenn, et al., Influence of thread design on pedicle screw fixation, J. Neurosurg. Spine 9 (1) (2008) 90–95.
[20] T.J. Choma, et al., Biomechanical analysis of pedicle screws in osteoporotic bone with bioactive cement augmentation using simulated in vivo multicomponent loading, Spine 36 (6) (2011) 454–462.

[21] A. Hashemi, D. Bednar, S. Ziada, Pullout strength of pedicle screws augmented with particulate calcium phosphate: an experimental study, Spine J. 9 (5) (2009) 404–410.
[22] T. Hasegawa, et al., Hydroxyapatite-coating of pedicle screws improves resistance against pull-out force in the osteoporotic canine lumbar spine model: a pilot study, Spine J. 5 (3) (2005) 239–243.
[23] B. Sandén, C. Olerud, S. Larsson, Hydroxyapatite coating enhances fixation of loaded pedicle screws: a mechanical in vivo study in sheep, Eur. Spine J. 10 (4) (2001) 334–339.
[24] J.N. Weinstein, B.L. Rydevik, W. Rauschning, Anatomic and technical considerations of pedicle screw fixation, Clin. Orthop. Relat. Res. (284) (1992)34-46.
[25] İ. Serkan, et al., Cortex of the pedicle of the vertebral arch. Part I: deformation characteristics during screw insertion, J. Neurosurg. Spine 7 (3) (2007) 341–346.
[26] B. Patel, et al., A novel route for processing cobalt–chromium–molybdenum orthopaedic alloys, J. R. Soc. Interface 7 (52) (2010) 1641–1645.
[27] T.C. Gascoyne, et al., In vivo wear performance of cobalt-chromium versus oxidized zirconium femoral total knee replacements, J. Arthroplast. 31 (1) (2016) 137–141.
[28] A.K. Madl, et al., Toxicology of wear particles of cobalt-chromium alloy metal-on-metal hip implants Part I: physicochemical properties in patient and simulator studies, Nanomedicine 11 (5) (2015) 1201–1215.
[29] A.K. Madl, et al., Toxicology of wear particles of cobalt-chromium alloy metal-on-metal hip implants Part II: importance of physicochemical properties and dose in animal and in vitro studies as a basis for risk assessment, Nanomedicine 11 (5) (2015) 1285–1298.
[30] S.C. Tan, et al., Tribocorrosion: ceramic and oxidized zirconium vs cobalt-chromium heads in total hip arthroplasty, J. Arthroplast. 31 (9) (2016) 2064–2071.
[31] J.-M. Brandt, et al., Performance assessment of femoral knee components made from cobalt–chromium alloy and oxidized zirconium, Knee 20 (6) (2013) 388–396.
[32] C. Chisholm, P.C. Poon, An in vivo kinematic study of the reverse shoulder joint replacement, Eur. J. Orthop. Surg. Traumatol. 22 (8) (2012) 655–660.
[33] J. Zhang, et al., Effect of implant shape and material properties on stresses in the glenoid components of total shoulder arthroplasties: A finite element analysis, in: ASME 2010 Summer Bioengineering Conference, American Society of Mechanical Engineers, 2010.
[34] G.B. Sharma, et al., Effects of glenoid prosthesis design variables on glenoid bone remodeling: A finite element based simulation and validation study, in: ASME 2008 Summer Bioengineering Conference, American Society of Mechanical Engineers, 2008.
[35] E. Witsø, et al., Improved comfort and function of arm prosthesis after implantation of a Humerus-T-Prosthesis in trans-humeral amputees, Prosthetics Orthot. Int. 30 (3) (2006) 270–278.
[36] J. Löhr, et al., Shoulder instability after primary shoulder joint replacement, Orthopade 27 (8) (1998) 571–575.
[37] A.A. Mohammed, S.P. Frostick, Linked shoulder replacement: current design problems and a new design proposal, Musculoskelet. Surg. 100 (1) (2016) 25–29.
[38] D.F. Gazielly, M.M. Scarlat, O. Verborgt, Long-term survival of the glenoid components in total shoulder replacement for arthritis, Int. Orthop. 39 (2) (2015) 285–289.
[39] G. Merolla, et al., Total shoulder replacement using a bone ingrowth central peg polyethylene glenoid component: a prospective clinical and computed tomography study with short-to mid-term follow-up, Int. Orthop. 40 (11) (2016) 2355–2363.
[40] D.Y. Chong, U.N. Hansen, A.A. Amis, Computational biomechanical analysis of fixation performance and bone resorption of tibial prosthesis implantation, in: ASME 2009 Summer Bioengineering Conference, American Society of Mechanical Engineers, 2009.

[41] Y.S. Lee, A.F. Chen, Managing bone loss in revision total knee arthroplasty, Ann. Joint 1 (2016) 17.
[42] D.Y. Chong, U.N. Hansen, A.A. Amis, The influence of tibial prosthesis design features on stresses related to aseptic loosening and stress shielding, J. Mech. Med. Biol. 11 (01) (2011) 55–72.
[43] B. McCormack, P. Prendergast, An analysis of crack propagation paths at implant/bone-cement interfaces, J. Biomech. Eng. 118 (4) (1996) 579–585.
[44] M. Viceconti, et al., Effect of the initial implant fitting on the predicted secondary stability of a cementless stem, Med. Biol. Eng. Comput. 42 (2) (2004) 222–229.
[45] ASM International, *http://www.asminternational.org/web/smst/newswire/-/journal_content/56/10180/26904243/NEWS*, 2017.

CHAPTER 4

# Osseointegration Concepts of Spinal Implants and Tissue Scaffolds

## 4.1 OSSEOINTEGRATION CONCEPTS OF SPINAL IMPLANTS

### 4.1.1 Introduction

Due to the variety of implants which are developed for the treatment of spinal injuries, spinal implants are recognized as the separate category of orthopedic implants. Spinal implants are implanted with higher risk severity and probability compared to other type of orthopedic implants due to the presence of spinal cord at the posterior aspect of the spinal column. Therefore, neurosurgery-spine surgeons prominently manage implantation of spinal implants. Associated injuries to the spinal column (which is treated by spinal implants) are due to degeneration, fracture, dislocation, or deformity of the vertebral or intervertebral segments. Nowadays, spinal implant developers are increasing due to development in surgical techniques, material engineering, advanced biomaterials, and clinical needs. Similar to other types of orthopedic implants, spinal implants are investigated in the various biomechanical, biological, and clinical aspects [1–18]. These aspects along with current concepts of osseointegration are discussed for spinal implants. Due to similar design concepts of pedicle screws with trauma screws, this type of spinal implant is reviewed in Chapter 3 and in this chapter; the intervertebral disc implants are focused.

### 4.1.2 General Aspects of Spinal Implants

Spinal implants have been manufactured from titanium alloys and polyetheretherketone (PEEK). Cobalt-chromium (CoCr) alloys have also been utilized in one type of spinal implants (disc replacements). Compared to other types of orthopedic implants, the variety of spinal implants is considerable and design conception in this area is greatly expansive. In view of biomechanical evaluation methods, ASTM F1717, ASTM F2077, and

*Osseointegration of Orthopaedic Implants*
https://doi.org/10.1016/B978-0-12-813384-2.00004-7

ASTM F2267 were normally addressed as the specific technical references for design validation of the spinal implants. In these standards, the only required loading conditions are compression and torsion loads while the bending moments of flexion-extension and lateral movements are physiologically crucial for the effective and reliable biomechanical evaluation of spinal implants. The protocols for carrying out of such biomechanical evaluations have been investigated and published as journal articles [3, 4, 15, 19–23] (clarification of the adequate and effective protocols would be out of the scope of this book).

### 4.1.3 Disordering or Injury of Intervertebral Disc

Majority of spinal injuries are affecting the intervertebral disc due to soft tissue structure of spinal disc, which is in the form of flexible tissue between the vertebrae. This flexible constitute is called intervertebral disc or interbody disc which is constructed from a fibrous connective tissue between the vertebrae (annulus) with gelatine form core (nucleus). Intervertebral disc would provide movement of the spinal column in multiple axes (flexion-extension rotation, axial rotation, lateral rotation, axial displacement, lateral displacement, anterior-posterior displacement). The range of motion of spinal column is strictly constrained with muscles, ligaments and intervertebral discs. Therefore, intervertebral disc would provide mobility and stability of the spinal column as the high fatigue load-bearing structure with a high risk of disorder or failure.

More importantly, neighboring of the disc with spinal cord would increase the severity of the disc disorder. In fact, deterioration of the biomechanical strength of the intervertebral disc causes an excessive compression of the disc from which the nucleus segment might be pushed out and suppressed spinal cord [diagnosis of the spinal cord irritation is implemented by magnetic resonance imaging (MRI) as shown in Fig. 4.1]. Suppression of spinal cord generates severe pain in spine and may in hand, leg, arm, foot, shoulder, etc. Biomechanical deterioration of disc is established in the long-term misuse or overloading of the spinal column in daily, work, or even professional sports activities. In other type of disc disorder, deterioration of the disc may cause over expansion or suppression of the disc from which the relative positioning of the adjacent vertebrae might be altered. This disordering effect may restrict the movement of the spinal column due to pain feeling because of spinal cord irritation.

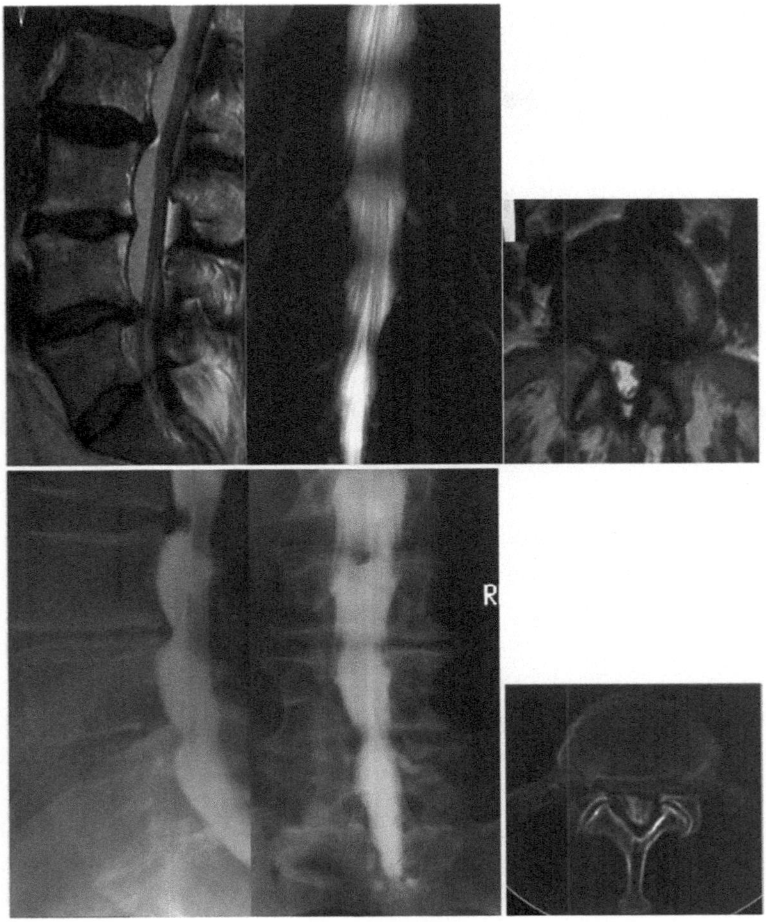


**Fig. 4.1** Irritation of the spinal cord in magnetic resonance imaging (MRI) scans. *(Courtesy of E. Shiban, M. von Lehe, M. Simon, H. Clusmann, P. Heinrich, F. Ringel, K. Wilhelm, H. Urbach, B. Meyer, M. Stoffel, Evaluation of degenerative disease of the lumbar spine: MR/MR myelography versus conventional myelography/post-myelography CT, Acta Neurochir. 158 (8) (2016) 1571–1578, and permission from Springer Link.)*

## 4.1.4 Treatment of Intervertebral Disc Deterioration or Degeneration

The treatment of disc degeneration or deterioration is managed through various surgical or nonsurgical methods. Nonsurgical methods are normally used in low- and mild-disc degeneration to reduce the internal pressure of the disc core (nucleus), from which the disc is repositioned in between of adjacent vertebrae with no irritation of spinal cord. Using of laser method for releasing of extended portion of the disc to the spinal cord and injection of Discogel (ethanol-based material) through the core segment of the disc tissue for reduction of core pressure are two alternative nonsurgical methods. Surgical methods are categorized according to the type of treatment and clinical demands which are currently organized as fusion and nonfusion treatment methods with different risk management. In general, spinal implants could be categorized as fusion or nonfusion implantable systems. In the treatment of severe spinal injuries in vertebral and intervertebral segments, fusion of intact neighboring or adjacent vertebral bodies to the injured zone or replacement of artificial disc prostheses might be managed.

### 4.1.5 Fusion and Nonfusion Treatments of Disc Degeneration

Fusion treatment is the formation of bone tissue in between of adjacent vertebrae of the degenerated disc to fill the space of intervertebral disc to form an integrated vertebral body. On the other hand, nonfusion treatment is replacement of an artificial prosthesis at the space of degenerated disc to simulate physiological ranges of motion of spinal column at the injured area without fully restriction of degrees of freedom (it is worth to mention that, in both fusion and nonfusion treatment methods, the degenerated disc is removed). Due to high level of physiological loading conditions at lumbar segment of spinal column, fusion treatment is more demanded while in cervical segment, nonfusion treatment is utilized in higher number of cases compared to lumbar segment. This might be due to a greater level of motion ranges in cervical segment of spinal column, which would be a considerable clinical demand in affected patients.

### 4.1.6 Development Concepts or Principles of Spinal Fusion Treatment

Implants for the fusion treatment of spinal injuries are developed based on some general biomechanical, biological, and clinical concepts. First, intervertebral disc height is essential to be preserved after implantation. Second, due to the extraction of the degenerated disc, the stability of spinal column at the operated spinal levels needs to be importantly provided. Third, the formation of new bone tissue in space of extracted intervertebral disc is significantly necessary to be conducted during the healing period. Forth, the mechanical strength of the implant is highly beneficial to be in optimum level [stiff enough to withstand physiological loading conditions and flexible enough to provide optimum mechanical stimulation for an effective and a long-term bone remodeling process (or bone formation-resorption cycle) at the injured zone]. Fifth, implant is strongly prohibited to have features with the risk of soft tissue irritation (which is indeed a general principle of orthopedic implants design concept). The implant strength must be sufficiently high against of physiological loading conditions (axial compression, flexion-extension bending, axial torsion, lateral bending) in both static and dynamic forms. Sixth, the subsidence of the implanted vertebrae should be negligible and not to affect the natural performance of attached muscles and ligaments to the spinal column. Seventh, dislocation or migration of the implant is greatly undesired and its effect on the clinical outcomes is significant. Thus, the fixation stability of the spinal implants is rigorously investigated in development stage through adequate biomechanical evaluation methods either computationally or experimentally. Eighth, anatomical form of the spinal column in healthy spine is mainly managed during implantation and instrumentation of the spinal fusion implants (Fig. 4.2 illustrates some designs of intervertebral fusion cages). In fact, the implant must provide anatomical stability of the spinal column at the injured or operated level. In this regard, various anatomy parameters of spinal column are focused such as disc height, lordosis angle, Cob angle, etc. These main design concepts are normally contemplated in the development of spinal fusion implants. Some of these concepts might not be well-intentioned in the past; however, well-established manufactures and developers are considering the reviewed eight development concepts or principles for the enhanced delivering of the various categorizes of spinal implants.

### 4.1.7 Development Concepts or Principles of Spinal Nonfusion Treatment

In addition of the expressed eight concepts for fusion spinal implants (Section 4.1.6), there would be other principles or design inputs that are essentially focused for the development of nonfusion spinal implants. One concept is the preservation of motion ranges of spinal column at the operated spinal level. The other concept is a strong attachment or the integration of the implant to the endplates with high shear strength against of multiple loading conditions. As the third specific requirement for nonfusion implant, low rate of wear at the contact surface of articulation components is examined during development phase. Furthermore, effective preservation of anatomical constraints is greatly beneficial in the enhancement of reliable ranges of motion and load transferring at the operated spinal level. Nowadays, the variety of spinal implants is interestingly growing at which new expandable

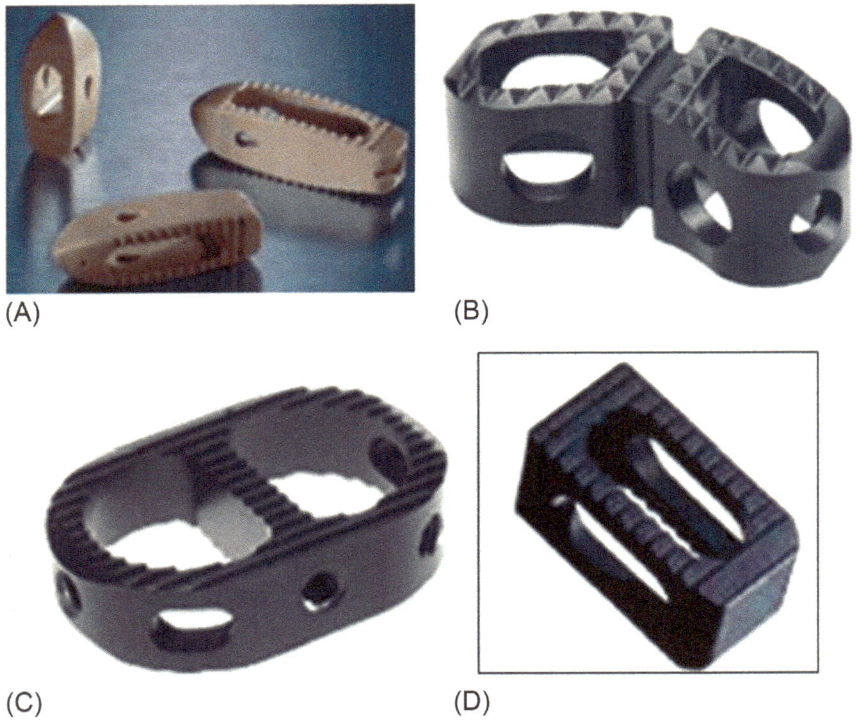


**Fig. 4.2** Various types of intervertebral fusion cages (A–D). Intervertebral fusion cages have been developed from various types of materials (Titanium alloys, 3D printed titanium, polyetheretherketone, bioactive ceramics, etc.). Likewise, these implants were manufactured with different design features and benefits to provide wider range of treatment techniques. Intervertebral fusion cages are enhanced with bone graft spacing for increasing of the compact structural bone in fusion of adjacent vertebrae to the operated intervertebral disc. *(Courtesy of V. Palepu, M. Kodigudla, V.K. Goel, Biomechanics of disc degeneration, Adv. Orthoped. 2012 (2012), Article ID 726210, and Hindawi.)*

fusion cages and 6-degree freedom nonfusion implants are more attractive for manufacturers as two superior implants with multiple biomechanical and clinical advantages. Some types of nonfusion implants are displayed in Figs. 4.3–4.5.

### 4.1.8 Osseointegration of Spinal Implants

From the design and development principles of fusion and nonfusion spinal implants, it could be concluded that osseointegration advantages of the implant is crucial for successful and effective clinical outcomes, particularly in the long-term treatment of spinal severe injuries. It could be proved that

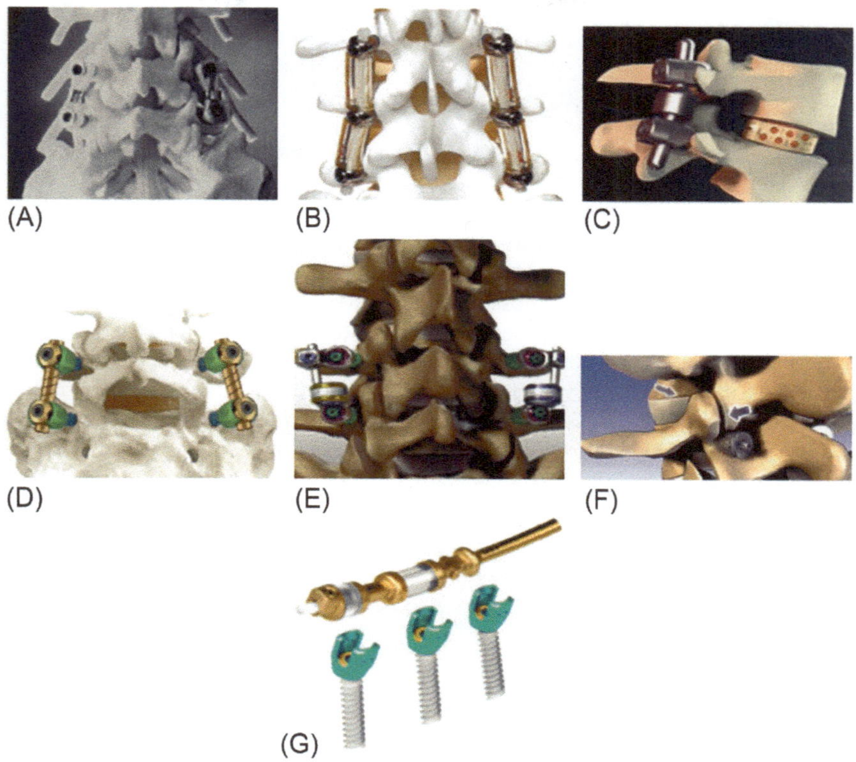


**Fig. 4.3** Various types of dynamic stabilization systems for fixation of spinal column at the affected levels (A–G). In majority of stabilization fixation, rigid spinal screw-rod systems are utilized, however, in some cases, dynamic rod-screw systems might be advantageous to reduce the rigidity of the spinal stabilization. In such cases, the dynamic element of the system could be developed through rod or screw. Dynamic screw and dynamic rods would facilitate local mobilization of the spinal column at the implanted level. Dynamic stabilization systems could be utilized either fusion or nonfusion implants. *(Courtesy of H. Serhan, D. Mhatre, H. Defossez, C.M. Bono, Motion-preserving technologies for degenerative lumbar spine: the past, present, and future horizons, Int. J. Spine Surg. 5 (3) (2011) 75–89, and permission from Elsevier, doi: https://doi.org/10.1016/j.esas.2011.05.001.)*

in majority of spinal implantations, integration of the implant to the vertebral is the advantageous target of the treatment procedure. Various portions of vertebral are targeted for the final mechanical and biological bone-implant integration. In pedicle screw insertion, integration or tapping of proximal portion of the screw shaft is highly essential to be mechanically achieved in cortical layer of pedicle bone at the posterior section of vertebral bone. Likewise, it is needful to achieve high shear strength of pedicle screw-bone integration at the distal portion of the screw in contact with the cancellous

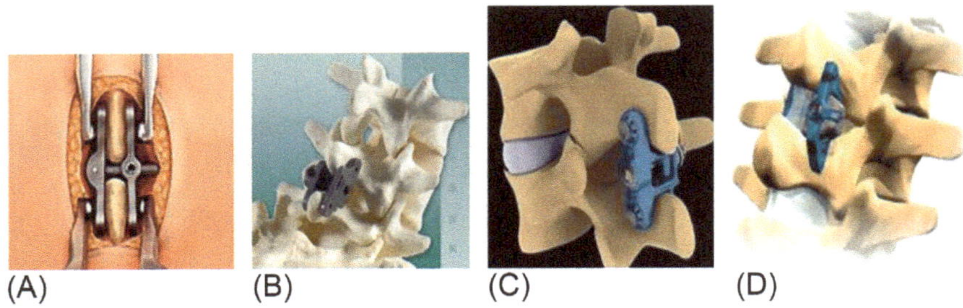


**Fig. 4.4** Various types of interspinous implants categorized as nonfusion implants (A–D). In such implants, the intervertebral disc is preserved and by constraining of the spinal posterior column, the motion ranges of the spinal anterior column is provided. This type of nonfusion implant would reduce the risk of clinical complications that might be observed in other techniques of spinal implantation and instrumentation using other types of spinal implants. The indication of interspinous implants is used for cases with mild injury of the intervertebral disc. *(Courtesy of V. Palepu, M. Kodigudla, V.K. Goel, Biomechanics of disc degeneration, Adv. Orthoped. 2012 (2012), Article ID 726210, and Hindawi.)*

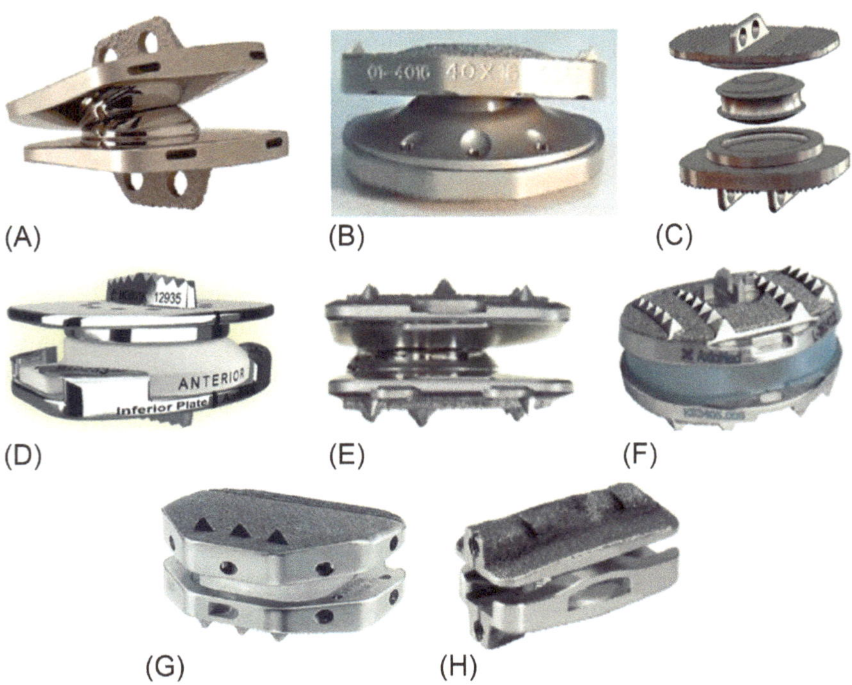


**Fig. 4.5** Various types of nonfusion artificial discs, also known as disc replacement or disc prosthesis (A–H). The main concept of such implant has been taken from total joint replacements which are used for the treatment of joint arthritis of knee, hip, shoulder, etc. Disc prostheses are normally manufactured from cobalt-chromium (CoCr) alloy and UHMWPE as the superior articulation biomaterials with low wear rate. Negligible wear rate would be achieved through cross-linking and antioxidant deposited UHMWPE and $ZrO_2$-coated CoCr alloys. *(Courtesy of H. Serhan, D. Mhatre, H. Defossez, C.M. Bono, Motion-preserving technologies for degenerative lumbar spine: the past, present, and future horizons, Int. J. Spine Surg. 5 (3) (2011) 75–89, and permission from Elsevier, doi: https://doi.org/10.1016/j.esas.2011.05.001.)*

bone. Biological integration of the pedicle screw at the distal shaft is highly advantageous in patients with a poor bone mineral density. Furthermore, intervertebral and vertebral fusion cages and intervertebral nonfusion prostheses are attached to the vertebral endplates, which necessitates early mechanical stability and long-term biological integration at the interface of implant and endplates. In interspinous implants, early and long-term mechanical integration and stability is highly demanded, however, biological integration with the lamina or spinous process of the vertebral bone might not be focused as the priority clinical demand.

### 4.1.9 High Advantageous of the Spinal Implant Osseointegration to the Vertebral Endplates

The critical osseointegration demand in spinal implants is currently the strong biological attachment of intervertebral/vertebral fusion cages and nonfusion disc prostheses to the endplates of vertebral bone. Reported migration of the implant in between of vertebrae has motivated researchers and manufactures to develop these implants with lower rate of migration or dislocation during treatment period. Likewise, it was highlighted that with the enhancement of osseointegration between the implant and vertebral endplates, lower amount of subsidence would be resulted. In fact, with the reduction of stress-shielding effect at the interface of vertebral bone-implant, the chance of long-term osseointegration is increased from which the greater biomechanical and clinical advantages of the implant are resulted. However, correlation between osseointegration and stress-shielding effect is related to mechanical aspects of bone-implant integration. From biological view, if the interface layer between implant and vertebral endplate would be formed with lamellar or mature bone tissue, it would act as a biological compressive supporter for the prevention of implant penetration through the endplate tissue.

### 4.1.10 Development of Intervertebral Fusion Cages in View of Osseointegration Aspect

Intervertebral fusion cages has been developed from aspects of material, design, manufacturing technology, surface coating, and surface porous layering. First, the design of titanium cages was enhanced with thinning of the cage walls to deduct the compressive strength of the cage. Next, PEEK polymer was introduced as a low-modulus and close mechanical strength to that of endplate tissue (modulus of 4 GPa for PEEK vs. 9 GPa for endplate). There was a challenging issue for the utilization of PEEK in spinal intervertebral/vertebral implants which were the compressive and flexural strengths.

Therefore, the first generation of PEEK cages was developed in bulky form, from which the cage was shaped as a spacer instead of a cage with a sufficient bone-grafting space. Afterwards, with improvement and progress in measurement of the physiological loading conditions which is transferred to the intervertebral disc (in multiple spinal range of motion in various axes), PEEK cage with thinner walls and higher bone grafting space was optimized (as the second generation) through the execution of biomechanical cadaver studies. Other than thinning of the cage walls, the shape of the cage was curved in contact surfaces with endplates to enhance distribution of compressive strength to higher area of bone-implant contact surfaces. The third generation of PEEK cages was introduced as the stand-alone fusion cages, which implanted without usage of a stabilization system such as screw-rod or plating system. Various types of stand-alone fusion PEEK cages have been developed recently (since last decade) in different designs and structures. The stand-alone PEEK cages are normally equipped with Ti blade, screw, or plate-screw. In parallel, static PEEK cages were enhanced to be expandable with higher clinical and instrumentation benefits. Likewise, other materials such as calcium phosphate-based bone graft, tantalum, three-dimensional (3D) printed Ti, biodegradable polymers, etc. were utilized in the development of intervertebral cages with advanced bioactivity performance compared to PEEK-based cages. However such investigated cages with new materials have not been commercialized globally. Bioactive PEEK composite could be considered as the fourth generation of PEEK cages. In this category of cages, the cage was developed from compounded PEEK polymer and nano-HA. HA-PEEK compound has been investigating for more than 10 years by biomaterial researchers [24–28]. HA-PEEK intervertebral cages are currently produced by a few companies in the United States and Europe, which have been implanted in the selected range of patients.

In order to enhance the osseointegration of the intervertebral cages, surface modification was focused as the fifth generation of the development progress. As reviewed in Chapter 7, surface nanostructuring has been found as a considerable osseoconductive way to promote the long-term integration of the bone to the implant. Coating of titanium on PEEK intervertebral cage and CoCr endplate component of disc prosthesis (artificial disc) have been developed for the promotion of implant bioactivity after implantation between the injured vertebrae. Ti coating is currently carried out through plasma spraying and sputter coating (magnetron DC or unbalance sputter) methods. It has been highlighted that a great adhesion strength of Ti-coating layer on the intervertebral cages are required for safe impaction of the cage during implantation [29]. Indeed, insufficient adhesion strength may cause detachment of the Ti particles from the coating layer, which increases the activity of macrophages cells for resorption of Ti particles. This would have undesired effect on bone remodeling process at the interface of bone-implant and osseointegration of the implant. Therefore, the higher effectiveness of sputtered Ti-coated layer with thickness of around 200–500 nm was reported compared to the thickness of plasma sprayed-coated layer in micro level [30].

## 4.2 TISSUE SCAFFOLDS

### 4.2.1 Introduction

Due to solid structure of the orthopedic implants, their integration to the bone (where required) would be with multiple challenges. The concept of porous structure in the development of implants was introduced to allow the growth of bone tissue through the structure of the implant from which in-depth osseointegration of the implant is perfectly achieved. However, porous structure would not have the mechanical strength as high as the solid structure at all. On the other hand, in majority of orthopedic implants, in-growth of bone formation is not desired or preferred. For instance, trauma plating systems (plate and screw) might be removed after the completion of fracture healing and bone growth through the plate would prevent the implant removal. Furthermore, if bone could grow inside the plate, abundant bone tissue is formed which is totally ignored. In fact, in some cases, the stability of the implant is highly associated to the integrity and stiffness of the implant, whereas in other type of implant, the stability might be governed by bone in-growth into the implant. Therefore, majority of orthopedic implants that are developed for the treatment of bone injuries are made with solid structure without in-growth of the bone. Joint replacement (e.g., knee, hip, shoulder, elbow, ankle, wrist, spinal disc) and internal trauma implants (e.g., plating systems, nailing systems, pining, endosseous screw fixation, cannulated screw fixation) are all manufactured in solid structure to provide the high mechanical strength against physiological loading conditions. Although such solid structuring would affect bone remodeling at the interface of bone implant, their usage is somehow inevitable based on the current available science and technologies. Biologic processing of bone remodeling is affected by stress shielding of the bone tissue which is caused by solid metallic implants (shielding of required mechanical stimulation for proliferation, mineralization, and maturation of osteogenic cells and expression of osteoblastic genes would cause ineffective processing of the bone remodeling). This effect would result in poor vascularization and mineralization of the collagenous tissues and thus poor on-growth of bone tissue on the implant surface. Today, enhancing the osseointegration of solid structural orthopedic implants is greatly under interest due to long-term advantages of biological integration of the implant to the bone. In other words, it has been revealed that the high mechanical strength of the implant could not guarantee the effective and successful clinical outcomes. In fact, due to variation in bone anatomy, bone quality, surgical techniques, severity of the bone injuries, risk of infection, postoperative treatment plan or exercises, patient willing to cooperate in the follow-up process, etc., the treatment of bone injuries is a complicated concern to achieve the best outcomes. Previously, clinical demands were in the low level and were mostly defined to fulfil the clinical requirements of the elderly people who has low demand of body movement in normal daily activities. This is whereas, nowadays, bone injures are occurred in young and mid-age population due to accident, malnutrition, static life style, etc., from which clinical demands are significantly higher than what were in the past. It could be concluded that rather than great mechanical stability, biological integration would be the key factor of orthopedic implant development. Since the growth of bone tissue is influenced by many factors, scaffold structuring implants could not be yet suggested in the treatment of bone tissue which is known as hard musculoskeletal tissue. However, numerous investigations [31–42] are currently carrying out to establish the usage of scaffold structure implants for the treatment of soft musculoskeletal tissues such as bone marrow, cartilage, spinal disc, cranial, retina, cornea, etc. Such implants are used for repairing or regeneration of the injured soft tissues with the low level of severity, which would not bypass their integration with the scaffold. These implants are called tissue engineering scaffolds which are reviewed in general concepts in this chapter. Since the trabecular bone is much porous than cortical bone, the combination of porous and solid structure in one integrated implant could be developed in the future.

### 4.2.2 Shaping of Tissue Engineering Scaffolds

In general, in science of tissue engineering, micro-architectural shaping of the scaffold is the key factor for the effective flowing of body fluid into the scaffold while the sufficient mechanical strength of such microstructure is a crucial requirement of longevity of the tissue scaffolds [43]. In fact, the size of pores and volume fraction of the scaffold architecture would be the challenging point to achieve the successful proliferation, differentiation, and ultimately formation of the targeted tissue into the scaffolds. The scaffolds are made of biodegradable materials to allow the complete formation of the tissue at the implanted zone.

### 4.2.3 Tissue Engineering Scaffold as the Regenerative Solution

Tissue engineering techniques would enhance the development of regenerative implants for repairing defective bone and soft tissues. In the treatment of bone injuries, although tissue engineered implants could not be a complete solution, it could be an advanced solution for the treatment of connective soft tissues to the bone (e.g., cartilage, spinal disc, bone marrow) to prevent propagation or advancement of the injury in the long-term use, such as what is observed in the degeneration of cartilage at the joints. Likewise, tissue engineered implant could be a supplementary solution in combination with solid orthopedic implants (e.g., trauma plating systems, joint replacement) to enhance the regeneration of the injured tissue. In fact, the stabilization of defected bone or bone fragments is managed by solid implants while tissue engineered implants are used to accelerate formation of the intended tissue. In cases, such as intervertebral fusion cages, the concept of porous structuring is currently encouraged for the treatment of spinal disc injuries (e.g., disc degeneration, herniation, or hydration). The concept of porous disc has been addressed for engineering of the spinal disc tissue [44–47] at which the adjacent vertebrae are hold by screw-rod stabilization system while disc tissue scaffold is replaced by the injured disc to establish a biological repair of spinal disc tissue (Fig. 4.6). In other case, tissue repair of knee cartilage at the initial stage could prevent severe degeneration of the cartilage tissue at the knee joint. With choosing of a suitable biomaterial (e.g., PEEK) with adequate mechanical strength, tissue scaffold of knee cartilage could be developed and inserted at the injured area to be well integrated with the surrounding cartilage tissue. Tissue engineering scaffolds could be developed for the treatment of severe bone loss or defect in long bones. As schematically shown in Fig. 4.7, by designing an adequate anatomical porous structure to be adopted with the defective area, the concept of spinal fusion process could be managed in the treatment of severe bone injuries in long bones (https://www.rmit.edu.au/news/all-news/2017/oct/just-in-time-3d-implants-set-to-transform-tumour-surgery-, shared on 30 Oct 2017 by James Giggacher).

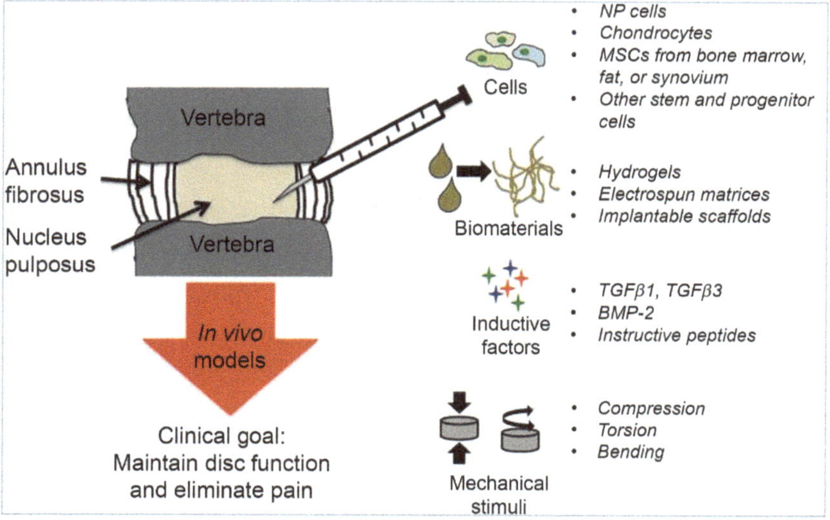


**Fig. 4.6** Schematic illustration of the biphasic spinal disc tissue scaffold for the treatment of disc injuries; the generative cells, growth factors, and proteins are deposited through the high porous tissue scaffold to build the new structure of the tissue at the implanted zone. The regenerated tissue would be investigated to have a sufficient biomechanical strength against compression, torsion, and bending stresses which are generated by the physiological motion of the spinal column in daily activities. *(Courtesy of G.D. O'Connell, J. Kent Leach, E.O. Klineberg, Tissue engineering a biological repair strategy for lumbar disc herniation, BioResearch (open access) 4(1) (2015) 431–445, and Mary Ann Liebert.)*

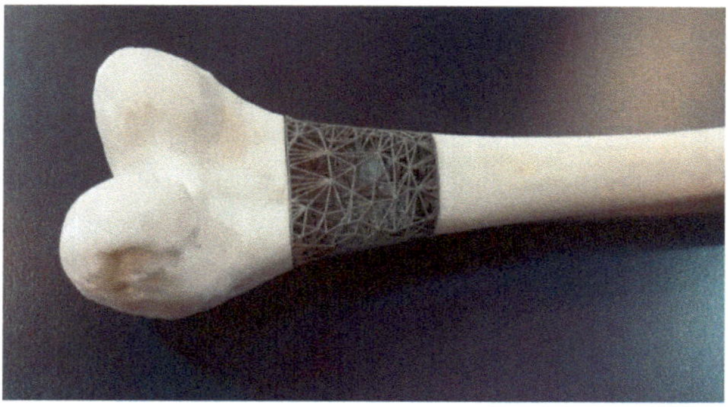

**Fig. 4.7** Schematic representation of the concept of scaffolding in the treatment of bone defect; the implantation of the high porous structure at the severe bone defect, particularly in long bones such as femur would be greatly advantageous for effective bone healing. By insertion of an advanced porous structural design at the defected zone, the bone formation would be canalized according to the bone anatomy, thus abundant bone formation is prevented. The designed porous structure is beneficial to be evaluated by computational and experimental fluid dynamic and solid structural methods to ensure its secure performance in the long-term implantation at the intended zone. *(Courtesy of RMIT University.)*

### 4.2.4 Culturing and Seeding of the Tissue Engineered Scaffolds

In general, tissue scaffolds are synthesized with similar architecture to the natural extracellular matrix of the targeted tissue [48] (tissue engineering scaffolds are fabricated through the process of 3D printing). In fact, the genesis cells of the tissue are essential to be able to generate the new cellular matrix in between of scaffold pores along with the formation of blood vessels for cell survival. The presence of growth factors or genes inside the scaffold for cell differentiation is highly important for the successful creation of tissue engineering scaffold. The targeted tissue cells and relevant growth factors are normally taken from the animals. For large bone defect, natural bone (autologous bone) transplants are beneficial to be utilized as it is being used in the treatment of bone fracture fixation or spinal vertebral fusion. In the case of bone fracture healing, the bone tissue might be dead in large scale at the fracture site, which is normally observed because of long duration from injury to the surgery or in open fractures. For the treatment of such bone fracture conditions, the large defected bone is completely removed from the body and autologous bone from iliac crest or fibula bone is extracted at the time of surgery to fill in the empty space of the defected bone. In other situation, the removed lamina bone from the back of vertebral bone is utilized as the autologous bone transplant or bone autograft to insert through the intervertebral fusion cage. However, in these two treatment cases, artificial bone graft could be utilized as well, if long surgical time would introduce serious clinical issues to the patient. Artificial bone graft could be fabricated as a normal porous or as the tissue scaffold, which is cultured by targeted cells, growth factors, alginate hydrogel, VEGF, proteins (e.g., BMP-2), etc.

### 4.2.5 3D Engineering of Tissue Engineered Scaffolds

3D printing process would facilitate controlled structuring of the scaffold in micro and macro level from various materials in powder form [49–54]. Indeed, this method constitutes the powder in the desired design even up to the internal zone. This advantage allows the formation of cavities from the top surface through the internal core of the scaffold in desired size. One of the main benefits of these cavities is guiding of angiogenesis and vasculogenesis cells through the whole parts of the scaffold to integrate the scaffold to the circulatory system of the body. In addition, the scaffold seeding or culturing at the specific zone becomes feasible for various biological and pharmaceutical factors such as extracellular proteins, growth factors, antibiotic, or antiinflammatory drugs [50, 55, 56].

## 4.3 SUMMARY

This chapter reviewed the spinal implants (in first part) and tissue engineering scaffolds (in second part) in general concepts to discuss the osseointegration aspects of these two categories of orthopedic implants (the other two categories "trauma fixation implants and total joint replacements" were reviewed in Chapter 3). Intervertebral disc has been found in the high risk of injury and thus majority of the spinal implants are developed for the treatment of intervertebral disc injuries or disorders. Intervertebral disc would be considered as the soft tissue which is connecting the vertebral bodies in spinal column. Intervertebral discs are bearing the various types of physiological loading conditions in six axes through which the central softer portion of the disc (nucleus) may transfer increased level of stress to the annulus fibrosus (strong outer ring of fibers to the nucleus)which is stiffer portion of the intervertebral disc. Annulus fibrosus might be weakened after long term from which the disc height is reduced and disc degeneration, herniation, hydration, or other type of severe disc diseases is occurred. The treatment of disc injuries would be managed by nonsurgical and surgical methods. Treatments methods by discogel drug, laser, magnetic resources energies, physiotherapy, etc. are noninvasive or nonsurgical methods that are utilized for the treatment of mild spinal disc injuries. For the treatment of severe injuries or disorders to the intervertebral disc, spinal implants are designed in two main categories of fusion and nonfusion implants. In majority of the surgical treatments, the spinal column is fixed at the affected levels along with the removal of spinal cord irritation. In such treatment, the implanted spinal levels might be fused or not. In fusion treatment, the spinal column might be fused from the posterior aspect where the lamina, spinous, or facet joint has been taken out. In some case, the severe injured intervertebral disc might be removed and interbody fusion cage implant would be replaced through the disc space. Nonfusion implants are the other surgical alternative for the treatment of intervertebral disc injuries. In such treatment methods, three main types of implants as dynamic screw-rod stabilization, interspinous stabilization, and disc replacement implants have been developed. In the view of osseointegration aspects, implantation of pedicle screws, interbody or vertebral fusion cages, and disc replacement prostheses would have some challenges. Similar to the other types of orthopedic implants, the long-term stability of these implants inside the bone tissue (cortical or trabecular bones) is crucial, particularly in low mineral density bones. Due to close application of pedicle screw to the trauma screws, osseointegration aspects of pedicle screws have been expressed in Chapter 3. As highlighted in Sections 4.1.7–4.1.9, integration of intervertebral implants (fusion cages and artificial disc replacements) with vertebral endplates would be the current development concept for the enhancement of bioactivity after implantation. Various types of surface processing methods (titanium coating, surface porous structuring) have been commercialized to promote the hydrophilic characteristics of the hydrophobic materials such as PEEK and CoCr alloys which are extensively utilized for the development of intervertebral fusion cages and disc replacement prosthesis, respectively. In Chapters 6 and 7, these processing methods are reviewed and effectiveness of each method in the enhancement of osseointegration of intervertebral implant is discussed. In the last section of this part (Section 4.1.9), the history of development of the intervertebral fusion cages was conceptually reviewed and the reasons of changes in material, design features/benefits, and even surface processing methods were discussed. In second part of this chapter, tissue engineering scaffolds have been reviewed in the general aspects of shaping architecture, regenerative solution, culturing and seeding, and 3D engineering. It has been highlighted that the highly porous structuring of these implants would allow regeneration of the soft tissues such as cartilage, spinal disc, bone marrow, etc. which might not be biologically achievable by human body regeneration process as what is observed in bone remodeling process. It is highly advantageous to incorporate the concept of porous scaffolding in the future development of solid structural orthopedic implants for the enhancement of bone-implant integration in the long-term implantation. In the following Chapters 5–9, this concept will be more discussed.

## 4.4 REMIND AND LEARN

**Q4.1** Express the indication of the spinal implants. What materials might be used for the development of spinal implants?

**Q4.2** What standards are currently utilized for the biomechanical evaluation of the spinal implants? (search for these standards and try to find out the main concepts of the testing methods) Why these standards would not be effectively usable for complete evaluation of the spinal implants? (consider the referred published articles to find out further biomechanical evaluation methods that might be applicable for spinal implants)

**Q4.3** Express the structure of the intervertebral disc and its function in mobility and stability of the spinal column.

**Q4.4** What effects might be caused due to the deterioration of biomechanical strength of the intervertebral disc? How is the spinal cord irritation diagnosed?

**Q4.5** What methods might be used to treat the disc degeneration or deterioration? Explain each method in general aspects.

**Q4.6** What is the definition of fusion and nonfusion treatment methods? How is the surgical demands of each method in lumbar and cervical columns?

**Q4.7** What are the development concepts of spinal fusion implants in view of biomechanical, biological, and clinical concepts (explain each concept)? Discuss how each development concept could enhance the safety and performance of the spinal implants.

**Q4.8** What technical concepts would be considered in the development of nonfusion implants? Express various types of nonfusion implants and compare their characteristics in the treatment of spinal injuries.

**Q4.9** What is the main aspect of spinal implant osseointegration? Which types of spinal implants are integrated through the vertebral bone? Discuss the integration characteristics of these implants to the vertebral bone.

**Q4.10** Discuss why osseointegration between the implant and vertebral endplates is crucial in implantation of intervertebral/vertebral fusion cages and nonfusion disc prostheses. Likewise, discuss how such critical osseointegration demand could be enhanced.

**Q4.11** From what general aspects, intervertebral fusion cages were developed? Describe the history of this development. Which methods of development are currently utilized as the superior methods?

**Q4.12** Discuss the advantages and disadvantages of solid and porous structural implants in the treatment of bone injuries.

**Q4.13** What are the parameter design of tissue engineering scaffolds? Express how each parameter would enhance the effectiveness of the scaffold in service.

**Q4.14** Explain how tissue engineering scaffolds would be utilized as the regenerative solution. What types of tissues are currently could be treated by tissue engineering scaffolds? How could severe bone loss or defect in long bones be treated by porous structural implant? Discuss the characteristics of such porous structural implant.

**Q4.15** What are the advantages of using 3D engineering methods in fabrication of tissue engineering scaffolds?

## 4.5 THINK AND DISCUSS

**Q4.16** Discuss what other kind of osseointegration concepts could be contemplated for the development of the spinal implants rather than what have been reviewed in this chapter.

**Q4.17** Intervertebral fusion cages are developed as the spacer to provide the stability of adjacent vertebrae during fusion and as the cage to facilitate the formation of the compact bone in between of adjacent vertebrae. Discuss how these two considerations are optimized and what effects would be introduced to osseointegration of the cages to the vertebral bodies.

## REFERENCE


[1] C.A. Chapman, Design of An Expandable Intervertebral Cage Utilizing Shape Memory Alloys, University of Toledo, 2011.

[2] L.S. Chatham, et al., Interbody spacer material properties and design conformity for reducing subsidence during lumbar interbody fusion, J. Biomech. Eng. 139 (5) (2017)051005.

[3] W.K. Cheng, S. İnceoğlu, Cortical and standard trajectory pedicle screw fixation techniques in stabilizing multisegment lumbar spine with low grade spondylolisthesis, Int. J. Spine Surg. 9 (2015) 46.

[4] S.-H. Chen, et al., Biomechanical effects of polyaxial pedicle screw fixation on the lumbosacral segments with an anterior interbody cage support, BMC Musculoskelet. Disord. 8 (1) (2007) 28.

[5] W.K. Cheng, et al., Pars and pedicle fracture and screw loosening associated with cortical bone trajectory: a case series and proposed mechanism through a cadaveric study, Spine J. 16 (2) (2016) e59–e65.

[6] J. Choi, S. Kim, D.-A. Shin, Biomechanical comparison of spinal fusion methods using interspinous process compressor and pedicle screw fixation system based on finite element method, J. Korean Neurosurg. Soc. 59 (2) (2016) 91.

[7] R.F. Heary, et al., Elastic modulus in the selection of interbody implants, J. Spine Surg. 3 (2) (2017) 163.

[8] F.Y. Hijji, et al., Post-market surveillance study of FLXfit™ TLIF interbody fusion device: protocol for a prospective case series study, Clin. Trials Degener. Dis. 1 (4) (2016) 181.

[9] J.C. Iatridis, et al., Role of biomechanics in intervertebral disc degeneration and regenerative therapies: what needs repairing in the disc and what are promising biomaterials for its repair? Spine J. 13 (3) (2013) 243–262.

[10] A. Kashkoush, et al., Evaluation of a hybrid dynamic stabilization and fusion system in the lumbar spine: a 10 year experience, Cureus 8 (6) (2016) e637.

[11] L. Marchi, et al., Radiographic and clinical evaluation of cage subsidence after standalone lateral interbody fusion, J. Neurosurg. Spine 19 (1) (2013) 110–118.

[12] A.G. Newcomb, et al., Stability with cortical screw-rod versus pedicle screw-rod fixation in the lumbar spine: effects of screw size, Spine J. 16 (10) (2016) S383–S384.

[13] V. Palepu, M. Kodigudla, V. Goel, Biomechanics of disc degeneration, Adv. Orthop. 2012 (2012)726210.

[14] P.J. Rao, et al., Spine interbody implants: material selection and modification, functionalization and bioactivation of surfaces to improve osseointegration, Orthop. Surg. 6 (2) (2014) 81–89.
[15] H. Schmidt, et al., The effect of different design concepts in lumbar total disc arthroplasty on the range of motion, facet joint forces and instantaneous center of rotation of a L4-5 segment, Eur. Spine J. 18 (11) (2009) 1695–1705.
[16] P. SLOSAR, Technological advancements in spinal fusion implants: a summary of the current scientific and clinical research on titanium engineered surfaces, J. Res. Found. 9 (1) (2014) 35–41.
[17] M. Stoffel, et al., Pedicle screw-based dynamic stabilization of the thoracolumbar spine with the Cosmic®-system: a prospective observation, Acta Neurochir. 152 (5) (2010) 835–843.
[18] F. Xavier, et al., Regional variations in shear strength and density of the human thoracic vertebral endplate and trabecular bone, Int. J. Spine Surg. 11 (2017) 7.
[19] Y. Nakajima, et al., Biomechanical analysis of a pedicle screw-rod system with a novel cross-link configuration, Asian Spine J. 10 (6) (2016) 993–999.
[20] R.P. Norton, et al., Biomechanical analysis of four-versus six-screw constructs for short-segment pedicle screw and rod instrumentation of unstable thoracolumbar fractures, Spine J. 14 (8) (2014) 1734–1739.
[21] C.-S. Chen, C.-H. Huang, S.-L. Shih, Biomechanical evaluation of a new pedicle screw-based posterior dynamic stabilization device (Awesome Rod System)-a finite element analysis, BMC Musculoskelet. Disord. 16 (1) (2015) 81.
[22] G.M. Mundis, et al., Contribution of round vs. rectangular expandable cage endcaps to spinal stability in a cadaveric corpectomy model, Int. J. Spine Surg. 9 (2015) 53.
[23] D.J. Cook, et al., Stability and load sharing characteristics of a posterior dynamic stabilization device, Int. J. Spine Surg. 9 (2015) 9.
[24] W. Lin, et al., Effect of preparation process on tensile strength of nano-hydroxyapatite/polyetheretherketone composite materials, in: TENCON 2015—2015 IEEE Region 10 Conference, 2015.
[25] R. Ma, et al., Research on the preparation of HA/PEEK gradient composites by impregnation method and the biosecurity, Mater. Sci. Forum 893 (2017) 35–42.
[26] M. Zhao, et al., Response of human osteoblast to n-HA/PEEK—quantitative proteomic study of bio-effects of nano-hydroxyapatite composite, Sci. Rep. 6 (2016) 22832.
[27] Y. Pan, J. Mao, J. Ding, Fatigue performance of hydroxyapatite filled polyetheretherketone functional gradient biocomposites, Mater. Technol. (2018) 1–8.
[28] M.R. Abdullah, et al., Processing of a multi-layer polyetheretherketone composite for use in acetabular cup prosthesis, J. Appl. Polym. Sci. 131 (20) (2014) 40915/1–40915/7.
[29] F.B. Torstrick, et al., Impaction durability of porous PEEK and titanium-coated PEEK interbody fusion devices, Spine J. 18 (5) (2018) 857–865.
[30] C.F. Michael Banghard, K. Silmy, A. Stett, V. Bucher, Plasma treatment on novel carbon fiber reinforced PEEK cages to enhance bioactivity, Curr. Dir. Biomed. Eng. 2 (1) (2016) 569–572.
[31] M. Fricia, et al., Osteointegration in custom-made porous hydroxyapatite cranial implants: from reconstructive surgery to regenerative medicine, World Neurosurg. **84** (2) (2015) 591. e11–591. e16.
[32] C. Rambo, et al., Osteointegration of poly-(3-hydroxybutyrate-co-3-hydroxyvalerate) scaffolds incorporated with violacein, Mater. Sci. Eng. C 32 (2) (2012) 385–389.
[33] S. Shi, et al., RhBMP-2 microspheres-loaded chitosan/collagen scaffold enhanced osseointegration: an experiment in dog, J. Biomater. Appl. 23 (4) (2009) 331–346.
[34] S.K.L. Levengood, et al., The effect of BMP-2 on micro-and macroscale osteointegration of biphasic calcium phosphate scaffolds with multiscale porosity, Acta Biomater. 6 (8) (2010) 3283–3291.

[35] X. Wang, et al., Topological design and additive manufacturing of porous metals for bone scaffolds and orthopaedic implants: a review, Biomaterials 83 (2016) 127–141.
[36] A. Palmquist, et al., Long-term biocompatibility and osseointegration of electron beam melted, free-form–fabricated solid and porous titanium alloy: experimental studies in sheep, J. Biomater. Appl. 27 (8) (2013) 1003–1016.
[37] T. Schouman, et al., Influence of the overall stiffness of a load-bearing porous titanium implant on bone ingrowth in critical-size mandibular bone defects in sheep, J. Mech. Behav. Biomed. Mater. 59 (2016) 484–496.
[38] A. Yánez, et al., Compressive behaviour of gyroid lattice structures for human cancellous bone implant applications, Mater. Sci. Eng. C 68 (2016) 445–448.
[39] J. Malmström, et al., Bone response inside free-form fabricated macroporous hydroxyapatite scaffolds with and without an open microporosity, Clin. Implant. Dent. Relat. Res. 9 (2) (2007) 79–88.
[40] K. Lee, et al., Bone regeneration via novel macroporous CPC scaffolds in critical-sized cranial defects in rats, Dent. Mater. 30 (7) (2014) e199–e207.
[41] T.H. Qazi, et al., Biomaterials based strategies for skeletal muscle tissue engineering: existing technologies and future trends, Biomaterials 53 (2015) 502–521.
[42] M. Nyan, et al., Accelerated and enhanced bone formation on novel simvastatin-loaded porous titanium oxide surfaces, Clin. Implant. Dent. Relat. Res. 16 (5) (2014) 675–683.
[43] M. Cavo, S. Scaglione, Scaffold microstructure effects on functional and mechanical performance: integration of theoretical and experimental approaches for bone tissue engineering applications, Mater. Sci. Eng. C 68 (2016) 872–879.
[44] G.D. O'Connell, J.K. Leach, E.O. Klineberg, Tissue engineering a biological repair strategy for lumbar disc herniation, Biores Open Access 4 (1) (2015) 431–445.
[45] B. Xu, et al., Intervertebral disc tissue engineering with natural extracellular matrix-derived biphasic composite scaffolds, PLoS One 10 (4) (2015)e0124774.
[46] A.T.H. Choy, B.P. Chan, A structurally and functionally biomimetic biphasic scaffold for intervertebral disc tissue engineering, PLoS One 10 (6) (2015)e0131827.
[47] Y. Wu, et al., A novel natural ECM-derived biphasic scaffold for intervertebral disc tissue engineering, Mater. Lett. 105 (2013) 102–105.
[48] C. Rücker, et al., Strategies and first advances in the development of prevascularized bone implants, Curr. Mol. Biol. Rep. 2 (3) (2016) 149–157.
[49] U. Gbureck, et al., Resorbable dicalcium phosphate bone substitutes prepared by 3D powder printing, Adv. Funct. Mater. 17 (18) (2007) 3940–3945.
[50] B. McAuslan, W. Reilly, Endothelial cell phagokinesis in response to specific metal ions, Exp. Cell Res. 130 (1) (1980) 147–157.
[51] A. Park, B. Wu, L.G. Griffith, Integration of surface modification and 3D fabrication techniques to prepare patterned poly (L-lactide) substrates allowing regionally selective cell adhesion, J. Biomater. Sci. Polym. Ed. 9 (2) (1998) 89–110.
[52] E. Sachs, et al., Three dimensional printing: rapid tooling and prototypes directly from a CAD model, J. Eng. Ind. 114 (4) (1992) 481–488.
[53] S.M. Allen, E.M. Sachs, Three-dimensional printing of metal parts for tooling and other applications, Met. Mater. Int. 6 (6) (2000) 589–594.
[54] M. Cima, et al., Three-Dimensional Printing Techniques, Google Patents, 1995.
[55] C. Kirkpatrick, et al., Endothelial cell cultures as a tool in biomaterial research, J. Mater. Sci. Mater. Med. 10 (10-11) (1999) 589–594.
[56] K. Whang, et al., Ectopic bone formation via rhBMP-2 delivery from porous bioabsorbable polymer scaffolds, J. Biomed. Mater. Res. A 42 (4) (1998) 491–499.

# CHAPTER 5

# Osseoconductivity Effects in Enhancement of Osseointegration

## 5.1 OSSEOCONDUCTIVITY ENHANCEMENT AT THE BONE-IMPLANT INTERFACE

### 5.1.1 Introduction

Orthopedic implants have been developed from various types of materials that could be compatible with human body. Majority of these biomaterials are biologically evaluated in aspects of toxicity, cytotoxicity, sensitization, etc. In fact, the inertness of utilized biomaterials for the development of orthopedic implants is highly desired in exposure of ionized body fluid. The chemical reaction of the implant material in vivo environment may cause release of implant particles through the body fluid and induce inappropriate response of body immune system. Therefore, various types of material enhancement techniques (e.g., coating, alloying, and antioxidation) are used to increase the corrosion resistance and inertness of the implant materials. Material such as stainless steel is enhanced by alloying with chromium to be corrosion resistant while titanium is anodized by formation of $TiO_2$ layer on its surface for the promotion of its biocompatibility and corrosion. Likewise, polymer such as ultrahigh-molecular-weight polyethylene (UHMWPE) is cross-linked for elimination of free molecular radicals. Although, inertness, corrosion resistance, and biocompatibility of the implant after implantation is vital, however, integration of the implant to the bone tissue is currently under interest to enhance biological, clinical, and even biomechanical benefits of the implants in treatment of bone injuries. In view of biological aspect, conductivity of the osteogenic cells to form bone tissue at the interface of the bone-implant would be an advantage. This could be achieved by development of such materials that can facilitate adsorption of osteoblast cells on their surface from which the formation of bone extracellular matrix (ECM) is resulted. Among the current biomaterials, titanium, tantalum, and magnesium alloys have the ability of osseoconductivity in contact with bone tissue. In bone remodeling process, activity of biological and mechanical factors is crucial for bone regeneration

*Osseointegration of Orthopaedic Implants*
https://doi.org/10.1016/B978-0-12-813384-2.00005-9

and resorption. Adhesion of osteoblast cells to the implant and its proliferation and differentiation is induced and conducted by appropriate bone-implant interaction in treatment of bone injuries. Involvement of growth factors, proteins, organic, and inorganic biological factors in bone generation would necessitate development of the orthopedic implant bioactivity for the enhancement of osseointegration and ultimately, satisfied clinical outcomes after long-term implantation. Loading of biological factors (e.g., growth factors, proteins, collagen fibers, and calcium phosphate) on the surface of bioinert implants would enhance induction of osteogenic cells to initiate bone generation. Due to the nanostructure of the bone tissue, nanostructured materials are being developed for the enhancement of bone formation at bone-implant interface in shorter period and higher integration strength while the density of new bone matrix is significantly increased.

Enhancement of osseoconductivity and osseoinductivity aspects of the bone-implant integration would enhance bone deposition, regeneration, or repair at the interface of the bone and implant. Disorders in the activity of osseoconductivity and osseoinductivity would enhance activity of the osteoclast and macrophage cells which results in excessive bone resorption. Osseoinduction is defined as inducing of biological factors that influence bone regeneration. In fact, osseoinduction stands on biological factors of bone remodeling process, whereas, osseoconduction is influenced by both mechanical and biological reactions to the surrounding environment of the bone cells. When an implant is implanted to the bone, bone remodeling process is highly affected. Therefore, in-depth understanding of biological and mechanical requirements of various stages of bone remodeling is necessary for developing the bioactivity of orthopedic implants. Body response or reaction to the implant is greatly affected by mechanical factors which influence the inductivity of osteogenic cells to form as mature bone matrix. Induction of osteogenic or osteoblast cells is initiated by creation of sufficient mechanical solid signals (as strain) which is then transmitted to the mechanical fluid pressure alteration followed by generation of electrical signals. This process would induce osteogenic cells to generate osteoblastic genes to induce the growth factors available in blood or body fluid for the creation of collagen around the osteoblast cells. It could be said that conduction and induction of osteoblast cells to create new lamellar bone are carried out simultaneously in bone remodeling process. One of the effective biological factors in bone regeneration is growth factors which are recognized as genes to facilitate expression and differentiation of osteoblast or osteoclast cells. All stages of bone remodeling process are carried out with the presence of growth factors. Proteins organize expression of growth factors and genes in the early stage of bone remodeling and absorption of minerals (e.g., calcium phosphate) to the fibrous collagen around the osteoblast cell. Mineralized osteoblast is formed as osteocyte cell and then matured to be structured as lamellar bone matrix.

In osseoinductivity enhancement, external loading of growth factors, platelet-rich plasma (PRP), bioactive mineral materials, and proteins at the interface of the bone-implant interface are implemented [1–3]. Due to the significant effect of the growth factors to promote regeneration of body tissues, PRP has been utilized as a carrier of growth factors for various treatment of body tissues including the associated tissues in orthopedics (e.g., bone, cartilage, muscle, tendon, and ligaments) [4–6]. Augmentation of PRP through the injured zone would increase concentration of platelets to increase the maturation rate of new cellular bone matrix (or increase of bone density), particularly, in treatment of bone defects [7]. Since the current challenge of orthopedic implant would be their integration with the bone tissue (osseointegration), bioactive ability or character of the implant is rigorously focused [8]. The focus is actually on the implant surface which is in contact with the bone. Modification of surface topography is associated to osseoconductivity enhancement of the implant. Processing of the surface morphology and establishment of the bioactive layer on the surface of the orthopedic implants would enhance conduction of the osteogenic cells, growth factors, proteins, and other biological factors through the implant surface to facilitate and accelerate bone formation on the implant surface. In order to enhance the bioactivity of the implant in contact surfaces with the bone, these two main strategies (surface morphology and bioactive surface layering) are processed as osseoconductive solutions. In this chapter, the general aspects of osseoconductivity are reviewed and further expression and discussion of the osseoconductive solutions are provided in Chapters 6 and 7.

### 5.1.2 Bone In-Growth and On-Growth for Enhancement of Osseoconductivity

Coating of bioactive materials on the surface of the implant is carried out to enhance the osseoconductivity of the implant through absorption of osteogenic cells (osteoblasts) and proteins on the surface of the implant. The bone in-growth and on-growth through the implant surface could enhance osseointegration of the implant. The porous coating of titanium with porosity in nanoscale (porosity with nano-pores) would have significant effect on the enhancement of implant-bone integration. Bone on-growth on the surface of the implant would not be a long-term solution due to deterioration of the bonding strength between the bone and implant from which the rate of bone resorption might be increased over the time. Normally, the bioactive coating (as a bone on-growth solution) is utilized to enhance the early bioactivity of the implant after implantation. The bioactive material is resorbed in long-term implantation, which may not accelerate formation of mature or lamellar bone matrix at the interface of the bone-implant. Alteration in morphology of the implant surface (e.g., roughening, blasting, and porous structuring,) would enhance the bone in-growth. The combination of bone on-growth and bone in-growth solutions (e.g., bioactive coating on blasting surface) has significant effect in enhancement of osseoconductivity at the bone-implant interface (Fig. 5.1). It is worth mentioning that the creation of inhomogeneous rough layer in macrolevel may not facilitate penetration of bone osteogenic cells through the implant surface and thus effective long-term osseointegration might not be achieved. In Chapter 7, porous structuring will be reviewed and discussed as the superior solution for enhancement of implant osseointegration in long-term implantation.

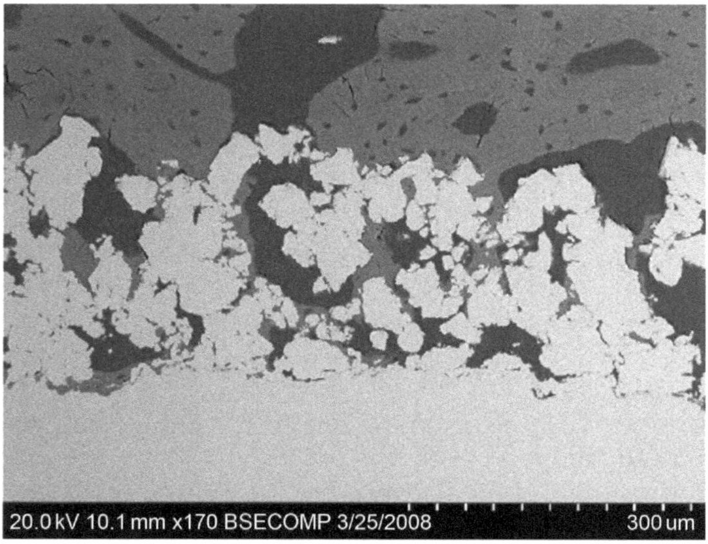


**Fig. 5.1** In-growth and on-growth of bone tissue inside and on the created porous layer on the surface of the implant, respectively; bone in-growth would have significant osseoconductivity effect and increase the longevity of the implant osseointegration. *(Courtesy of D. Chen, et al., Osseointegration of porous titanium implants with and without electrochemically deposited DCPD coating in an ovine model, J. Orthop. Surg. Res. 6(1) (2011) 56 and BioMed Central.)*

### 5.1.3 Osseoconductivity Rate in Cortical and Cancellous Bones

Osseoconductivity rate at the interface of the bone and the implant with porous surface could be evaluated through examination of the on-growth and in-growth rates of bone formation. The bone on-growth rate was increased in cortical bone at 12 weeks compared to 4 weeks of postimplantation, while it was decreased in cencellous bone. The reason might be early attachment of the high porous cancellous bone through the thin porous layer of implant due to a higher rate of mechanical stimulation for induction of the osteogenic cells compared to cortical bone. The interesting finding was that the bone on-growth rate at the interface of the cancellous bone and implant with porous layer is similar in either with or without bioactive coating layer. This would show the greater early attachment of the trabecular bone through the porous surface of the implant compared to the cortical bone. In cortical bone, the mechanical stimulation might not be in the sufficient level as effective as the trabecular bone at early implantation. After 12 weeks of implantation, the mineralization of the new formed tissue is gradually increased and thus the new formed bone matrix become more compact and can bear higher level of strain, which causes increasing of the bone on-growth rate in cortical bone and deduction of the bone on-growth rate in cancellous bone. Bone in-growth rate in cancellous bone is reduced from 4 to 12 weeks. Based on what Chen et al. [9] reported, the in-growth rate is increased until 4 weeks and then sharply dropped at 8 weeks and increased again at 12 weeks in cancellous bone either with and without bioactive coating implants. This irregular trend would not be discussed clearly, as multiple factors might be involved. In one aspect, creation of higher percentage of lamellar or mature bone at the porous layer would reduce the in-growth rate. In other aspect, the thickness of the porous layer might be reduced over the time due to its thin walls.

### 5.1.4 Osseoconductivity in Treatment of Bone Defects

In order to treat the bone defect, the synthetic bone graft is an alternative solution. The bone defect might be resulted from tumor, high trauma, severe infection, osteotomy treatment, or other surgical requirements. In fact, after removing of the defected bone, a large gap might be created between the intact bones from which bone healing would not be achieved. The autologous bone graft might be used intraoperatively in areas that have sufficient boundaries to hold the bone graft such as substitution of bone graft with nucleus portion of the intervertebral disc while annulus is preserved to keep the bone graft for spinal fusion. Bone autologous bone graft is also used to fill the space of bone loss in treatment of bone fractures when the space would be surrounded by the bone fragments. The xenologous bone grafts (which are taken from the other species with close bone mineral densities to the human bones [10]) are the other alternative to be used in treatment of bone loss or fusion cases which are formed in regular geometrical shapes such as cylinder or square. Allograft might be used for the treatment of large bone defect when the defect has affected a large portion of the bone which could not be treated by the autologous bone grafts due to several clinical and surgical limitations [11]. Other solution is to synthesize osseoconductive biomaterials that can conduct formation of bone ECM at the bone defect area (Fig. 5.2). These materials are synthesized from a similar mineral

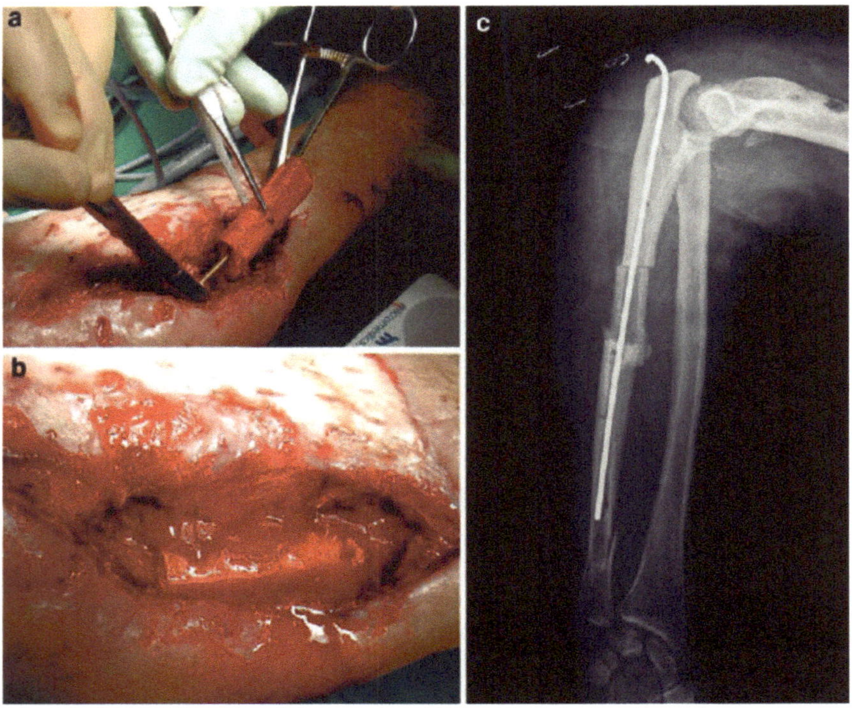


**Fig. 5.2** Implantation of the bone substitute for treatment of severe bone defect (A–C); the bone substitute is fabricated from calcium-phosphate-based materials to conduct the bone formation from the intact or nondefective bone through the substitute structure. The bone substitute could be loaded or enriched with platelet-rich plasma (PRP) as implemented in the utilized bone substitute in this figure. *(Courtesy of S. Paderni, S. Terzi, L. Amendola, Major bone defect treatment with an osteoconductive bone substitute, Musculoskelet. Surg. 93(2) (2009) 89–96 and get permission from Springer publisher.)*

constitutive of bone tissue. Materials such as tricalcium phosphate (TCP), calcium sulfate, bioglass, hydroxyapatite (HA), and biphasic calcium phosphate (BCP) are some of the synthetic materials that are currently used to fabricate bone substitute for the treatment of bone defect or losses [11]. Such bone mineral biomaterials would enhance adherence and mitigation of the protein and differentiation of the osteoblasts cells [12].

## 5.2 OSSEOCONDUCTIVITY ENHANCEMENT BY BIOACTIVE MATERIALS

### 5.2.1 Osseoconductive Advantageous of Nanostructured Calcium Phosphate Materials

In order to increase the effectiveness of osseoconductivity, utilization of similar nanostructured material with inorganic portion of the bone tissue would have conduction effect on formation of the new bone ECM [13–17]. The calcium-phosphate-based materials in nanoscale (1–100 nm) is in the similar scale of inorganic constitute of the bone tissue from which induction rate of various stages of bone remodeling is significantly enhanced. Bone tissue is a nanocomposite with organic and inorganic constitutes; the organic portion is collagen type I, which consists of nanofibers with diameter of 50–500 nm and the inorganic constitutes are structured with nonstoichiometric HA crystals with approximate length of 100 nm and width of 20–30 nm. It is worth mentioning that the thickness of HA crystals is about 3–6 nm which shows how small the inorganic phase of bone tissue is [18, 19]. Calcium phosphate is the main portion of the bone inorganic constitutes which is currently synthesized in various forms (Table 5.1).

**Table 5.1** Various types of synthesized calcium phosphates

| |
|---|
| Nano-calcium phosphate monohydrate |
| Monocalcium phosphate anhydrous |
| Dicalcium phosphate dehydrate |
| Dicalcium phosphate anlydrous |
| Octacalcium phosphate |
| α- and β-tricalcium phosphate (αTCP& βTCP) |
| Amorphous calcium phosphate (ACP) |
| Calcium-deficient hydroxyapatite |
| Hydroxyapatite (HA) |

### 5.2.2 Long-Term Effect of Nanostructured Calcium Phosphate Materials

One of the challenges of the osseoconductivity is its effectiveness in long-term use. In fact, with the forming of osseoinductive material (e.g., growth factors, calcium phosphate, and proteins) on the surface of the implant, conduction of bone formation on the surface of the implant is remarkably increased at the bone-implant interface at early implantation. However, the osseoinductive constitute might be resorbed or released gradually to the human body fluid, thus causes deduction in formation of the collagen fibers, mineralization of collagen, and maturation of the new bone matrix. On the other hand, adhesion strength of the osseoinductive constitutes or materials to the surface of the implant would be crucial to establish the presence of osseoinductive material for a longer period after the implantation. The main challenge in this topic is the temporary effect of osseoconductivity in enhancement of the implant osseointegration. Indeed, the existing osseoinductive material on the surface of the implant enhances various stages of bone remodeling process for the formation of bone matrix at the bone-implant interface. The optimum amount of osseoinductive material and how to process it to the implant surface would be controversial for an effective osseointegration of the implant at early, short-term, and long-term implantation and a better functionality of the implant in place. The penetration of the osseoinductive material through the surface of the implant is the current solution which is processed physically or chemically to form an adhesive layer of the osseoinductive material. Chemical bonding of the osseoinductive material to the implants [implants are normally made from stainless steel, cobalt chromium alloy, titanium alloy or polyetheretherketone (PEEK) polymer] might not be achieved effectively due to difficulty in conduction of adequate conditions for chemical reaction between the implant and osseoinductive material. Therefore, physical bonding methods (e.g., plasma spraying, dip coating, electrochemical deposition, vapour deposition, etc.) are currently utilized by manufacturers to load osseoinductive materials on the surface of the implant. Although, the deposition of osseoinductive material is being developed, invention of an osseoconductive structure that can be recharged with new osseoinductive particles seems to have advantages for long-term osseointegration as needed in total joint replacements for more than 20 years.

### 5.2.3 Nano-Synthesized Calcium Phosphates as Superior Osseoinductive Material

The nano-calcium phosphate (nano-CaP) has been found to have better physiological interaction with stem cell than the micro-size calcium phosphate [20, 21]. This would establish a greater biological response from the host bone to form a stronger adhesion of the bone tissue to the implant. The nano-CaP would enhance its attachment to the stem cells from which the loaded nanoCaP on the surface of the implant generates a physicochemical bonding or link between the implant surface and bone cellular matrix. It is worth mentioning that the presence of host bone tissue at the interface of the bone-implant with a gap of few micrometres could promote the creation of this physicochemical bond. In real scenario, the gap at the interface of the bone-implant may not be consistent. Therefore, initial bond between the CaP layer at the surface of the implant and host bone is not homogenous at the bone-implant contact area. From this view, placement of the implant (loaded with CaP) is one of the challenges of the surgical techniques. The nanostructured CaP may be synthesized or processed in three groups as nanostructured CaP cement (CPC), nanostructured CaP scaffold, and nanostructured CaP coating for the enhancement of implant bioactivity. The CPCs are used as synthetic bone graft (e.g., fusion of spinal vertebrae), bone augmentation (such as in implantation of cannulated pedicle screws), replacement of total joint prosthesis at the joints, and bone substitutes for the treatment of bone defects [22–24]. This group of nanostructured CaP has been initially developed from the tetracalcium phosphate (TTCP) with chemical formation of $Ca_4(PO_4)_2O$ and dicalcium phosphate anhydrous (DCPA) with the formation of $CaHPO_4$ [25]. The nano-bioactive bone cements have been utilized for the treatment of craniofacial defects [26] with approval of FDA (The US Food and Drug Administration) in 1996. This type of developed material is categorized as the cement constitute due to its self-curing or self-hardening in exposure of aqueous solution by which a compact layer of HA is formed at the target zone. Mechanical properties of the nano CPC have been enhanced by compounding with some biocompatible additives such as absorbable fibers [27] and chitosan [28]. The nanostructured CaP can be developed as composite to be used for the treatment of bone injuries in high load bearing zones (e.g., total knee replacement, fracture fixation of tibia, femur, or pelvic bones). The CaP composites are good alternative to fabricate scaffold or cage based implants such as tissue engineering scaffolds, spinal intervertebral cages, and bone substitutes. Furthermore, for the enhancement of the bioactivity of metallic or polymeric orthopedic implants the CaP particle in nanosize individually or in combination with other elements or compositions such as strontium, might be coated on the surface of the implant

### 5.2.4 Greater Adhesion of Nano-Amorphous Calcium Phosphate to Stem Cells

The interesting point in adhesion of stem cells to the nano-CaP is the existence of cytoplasmic extensions on mesenchymal stem cells (MSCs) which would enhance penetration of MSCs to the nano-pores of the CaP [29]. Ramalingan et al. [30] highlighted the effect of amorphous calcium phosphate nanoparticles (nACP) to promote adhesion and proliferation of osteoblast cells. They have developed a nano-fibrous scaffold with nano-powder of CaP using two-spinnerette electropolishing. It was shown that due to the formation of the gradient regions in amorphous structuring of CaP, trapping of higher amount of osteoblast cells is resulted and therefore greater interaction between the stem cells and the CaP amorphous structure would be achieved. Orientation of cell adhesion to the implant would be an advantageous matter that can enhance homogeneity and adhesion strength of the stem cells [31, 32]. Compounding of CaP with fibrous materials such as collagen or PLGA with a fiber diameter of 100–350 nm (as developed by Jose et al. [33]) could enhance the creation of nano-grooves or other type of nano irregularities (e.g., pits, wells, steps, pillars, pores, and ridges). Such CaP composites would facilitate better regulation of the stem cell orientation on CaP morphology.

## 5.3 OSSEOCONDUCTIVITY ENHANCEMENT BY COMBINED BIOACTIVE-POROUS SURFACE LAYERING

### 5.3.1 Combination of Surface Topography and Bioactive Layering for Long-Term Osseoconductivity

In majority of the surface enhancement for greater osseointegration effect, early bone-implant integration is obtained which is valuable for an early stability of implantation system; however, long-term osseointegration of the implant needs further investigation. As a solution, a combination of enhancement in surface topography and surface loading with bioactive agent could be developed for an early and long-term osseointegration with a high bone-implant interfacial strength. Osseoconductivity is effectively achieved when osteogenic cells have the chance of adsorbing on the surface of implant and this could be obtained by rough topography of the surface (Fig. 5.3) and presence of osseoconductive materials such as, Mg and potassium. In one study, the Mg-loaded implant was compared with laser-etched implants [34]. The implant with laser-etched surface showed higher interfacial strength than the Mg-loaded implant. This might be due to the greater

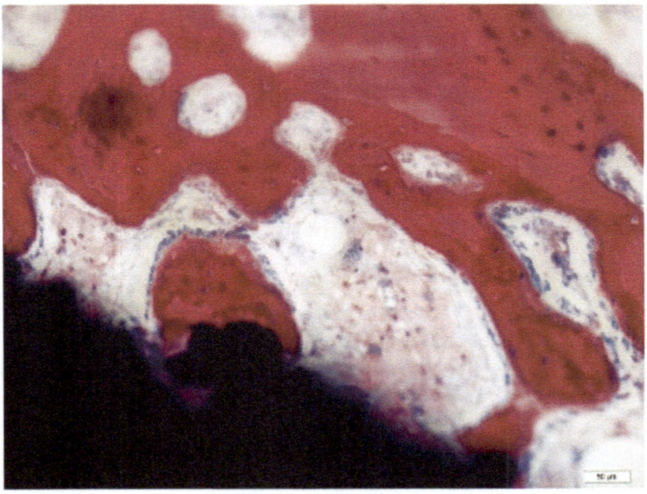

**Fig. 5.3** Illustration of the osteoblasts (black dots) and new lamellar bone (red color) on the rough surface of the implant after 2 weeks of implantation; this would show active osteogenesis on the rough surface. *(Courtesy of D. Chen, et al., Osseointegration of porous titanium implants with and without electrochemically deposited DCPD coating in an ovine model, J. Orthop. Surg. Res. 6(1) (2011) 56 and BioMed Central.)*

roughness of the surface in laser-etched surface treatment. In fact, enhancement of surface bioactivity and surface topography are two strategies in the development of osseointegration. In view of osseoconductivty improvement, the presence of an osseoconductive agent such as Mg, conducts osteogenic cells to be located on the surface of the implant and enhance formation of bone ECM. However, such degradable osseoconductive materials (e.g., Mg) are resorbed in exposure of body fluid and thus they should be trapped or deposited through the surface of the implants. Even with an increase in the thickness of loaded or coated layer of the bioactive material on the surface of the implant, long-term osseointegration would not be effectively achieved [35].

## 5.3.2 Mesoporous Layer With Nano-Pores as the Osseoconductive Layer

Formation of a porous layer with nano-pore size has been significantly effective in enhancement of osseointegration. When the pore size of the porous structure is in the range of 2–50 nm, it is known as mesoporous structure. The mesoporous layer of $TiO_2$ was developed on the surface of the titanium implant using evaporation-induced self-assembly method [36]. This nanopore layer was created as the host layer to allow loading of various types of drugs [37], calcium phosphate [38–43], proteins (growth factors), etc. that can promote osseoconductivity, mineralization, osseoinductivity, and ultimately, bone apposition through the surface of the implant [44–47]. The enhanced osseoconductive implants would allow better early integration of the implant from which the implantation positioning of the implant could be effectively retained against that of physiological loading conditions (physiological loads may migrate or displace the implant from the original implantation position).

### 5.3.3 Mg-Deposited Mesoporous Layer With Enhanced Osseoconductivity

Magnesium (Mg) has a key role in the regulation of biochemical activity of the bone formation process. The Mg has significant influence on proliferation of the osteoblast cells (Mg would increase the level of nitric oxide from which the activity of osteoclast cells is induced [48, 49]). The atomic concentration of the Mg has been altered by researchers to enhance the effectiveness of Mg osseoconductivity [45, 50–52]. It was observed that atomic concentration of 8%–9% [36, 45, 52] would result in greater osseointegration of the loaded Mg implants. The loading of magnesium on mesoporous layer was performed by Galli et al. [36] as an osseoconductive material to investigate the early osseointegration of the implant in rabbit bone after 3 weeks. In their research, a mesoporous layer was developed on titanium implant with nano-pores to enhance adsorption of inorganic ions through the surface of the implant. The aim was to establish an osseoconductive layer to trap the nano osteogenic cells (osteoblast cells) and facilitate their migration and differentiation on the surface of the implant [37, 53, 54]. To achieve that, a homogenous coating or establishment of this mesoporous layer on the contact interface of the bone-implant would be beneficial for the effective osseoconductivity of the implant. Although Galli et al. [36] reported their result in 2015 for early implantation, however, their solution as creation of nano-pores $TiO_2$ as the surface topography enhancement and loading of Mg through the nano-pores as the surface bioactive agent, would be a valuable improvement in the enhancement of long-term osseointegration of orthopedic implants.

### 5.3.4 Enhanced Osseoconductivity by Deposition of Bioactive Material on Titanium Porous Layer

Coating of Ti on the implant surface would be the other solution, which is normally processed by the plasma spraying method. The titanium coating with nano-pores would increase the on-growth and in-growth formation of lamellar bone on the surface of the implant. The osseoconductivity of the titanium porous coating could be promoted by additional coating of the bioactive material (e.g., bioactive ceramics, Mg, chitosan, etc.). If the topography of titanium porous layer in nanoscale, the bioactive particles might not be penetrated through the nano-pores. Thus, the bioactive coating layer would not be effectively integrated with the nano porous Ti layer. Although the bioactive layer would enhance the early bioactivity of the implant, however, it would be resorbed over the time and thus the bone in-growth might not be achieved. On the other hand, the nanostructure of the porous layer might be affected by re-coating of the bioactive material on the porous layer. To overcome this issue, development of a controlled and noncompromised coating process to preserve the nanostructure of the created porous layer was developed by Chen et al. to deposit the bioactive material through the Ti porous layer [9]. They have utilized the electrochemical cathodic deposition process [55, 56] to coat a thin and homogenous layer of calcium phosphate on the titanium porous surface with nano-pores. In their results, the increased rate of bone in-growth was observed after 4 and 12 weeks of postimplantation.

## 5.4 SUMMARY

Various aspects of osseoconductivity and osseoinductivity at the interface of the bone and the implant have been reviewed and discussed in this chapter. In first part, the osseoconductive solutions for development of the orthopedic implants have been addressed. In general, the current solutions could be categorized in two main groups; one is creation or establishment of the porous layer on the surface of the implant, and the other is formation or coating of the bioactive material on the implant surface. The latter group has been commercially developed for bioactivity enhancement of the orthopedic implants such as coating of HA on the implant surface, while the former group is going to be the superior solution for the promotion of the bone-implant integration in orthopedic implant industry such as forming of the tantalum porous layer on the femoral knee component. As highlighted in the reviewed literature in this part, the advanced solution would be the combination of both groups to establish the osseoconductive layer on the surface of the orthopedic implant. In fact, the porous layer would act as the host layer for the formation of extracellular bone matrix through the implant surface and thus increasing of the osseointegration longevity. The main challenge of porous layering on the surface of the implant is the creation of porous layer with effective porosity characteristics such as pore size, pore interconnection, thickness, porous layer strength, and porous layer adhesion strength to the implant. All these effective characteristics will be discussed in Chapter 7. As generally reviewed in Sections 5.3.2–5.3.4, the porous layer with nano-pores (mesoporous layer) has been found significantly effective for adhesion of the osteogenic cells on the surface of the implant. However, there would be some challenging aspects (e.g., porous layer strength and coating of bioactive material in nano powder form) that necessitate further investigations to introduce mesoporous layer as an effective solution for the enhancement of implant osseointegration. In osseoconductivity, conduction of bone formation on and inside the pores of the porous layer was expressed as bone on-growth and in-growth respectively. It was highlighted that both osseoconductivity types could be beneficially contemplated for the long-term integration of the bone tissue to the implant. After implantation, the bone on-growth rate would be first increased at the interface of the implant and the cancellous bone and then decreased, while it would be decreased and then increased at the implant interface with the cortical bone. Likewise, the bone in-growth rate is increased in both cortical and cancellous bone at early implantation period. After couple of weeks (e.g., 4 weeks), the in-growth rate has a sharp drop in the cancellous bone. The other reviewed aspect of osseoconductivity was the conductivity of the bone formation through the bioactive bone substitutes in treatment of bone defects or loss which is obtained by the calcium-phosphate-based materials (the function of this type of biomaterials for conduction of the bone formation is reviewed in Chapter 6).

## 5.5 REMIND AND LEARN

**Q5.1** Express the various aspects of biocompatibility and bioactivity of the orthopedic implants. Provide some examples for enhancement of implant bioinertness in human body fluid.

**Q5.2** What is the definition of osseoconductivity and osseoinductivity? Discuss the induction and conduction of the osteogenic cells to be matured as lamellar bone in bone remodeling process.

**Q5.3** How could osseoinductivity be improved at the bone-implant interface? What types of biological factors might be applicable for osseoconductivity enhancement?

**Q5.4** How would bioactivity of the implant be enhanced? Would it be associated to osseoconductivity or osseoinductivity, why? What are two main strategies of bioactivity enhancement in the orthopedic implants?

**Q5.5** What are the benefits of bone in-growth for enhancement of osseointegration of the orthopedic implants? How would bone in-growth be improved?

**Q5.6** How would osseoconductivity of the implant with porous surface be evaluated in cortical and cancellous bones? Provide the feasible reasons for increase or reduction of the osseoconductivity rate in cortical and cancellous bone after implantation period.

**Q5.7** Describe the osseoconductivity concept treatment of bone defect using bone grafts.

**Q5.8** What is the osseoconductive advantageous of nanostructured calcium phosphate materials? Why nanostructured calcium phosphate materials have superior osseoconductivity compared to micro or macro calcium phosphate materials? Discuss your answer using the sizes of the bone organic and inorganic constitutes.

**Q5.9** What are the challenges of forming of the osseoinductive materials (e.g., growth factors, calcium phosphate, and proteins) on the surface of the implant for long-term effectiveness? Discuss how the long-term effectiveness of the osseoinductive materials could be enhanced. Which type of chemical or physical bonding might be developed to form the osseoinductive materials on the implant surface?

**Q5.10** What types of nano-synthesized calcium phosphates might be developed for the treatment of bone injuries? What are the benefits of nano-synthesized calcium phosphates compared to micro size calcium phosphates?

**Q5.11** Discuss the adhesion of nano-amorphous calcium phosphate to stem cells. How would regulation of stem cell be improved on CaP morphology?

**Q5.12** Discuss the effect of surface topography and bioactive layering on long-term osseoconductivity. What is the long-term effectiveness of bioactive-coated material on the smooth surface? Does higher thickness of bioactive coating layer enhance the longevity of the osseointegration?

**Q5.13** How could the mesoporous $TiO_2$ layer be created on the surface of the titanium implants? What are the benefits of such mesoporous layering?

**Q5.14** Express the bioactivity or osseoconductivity effects of magnesium? What are the osseoconductivity advantages of Mg deposition on mesoporous $TiO_2$ layering?

**Q5.15** Discuss the enhancement of the osseoconductivity by deposition of bioactive material on titanium porous layer. What are the challenges of the deposition of bioactive material on titanium nano-porous layer? What processing method has been developed for controlled and noncompromised coating of the bioactive materials on titanium nano-porous layer?

## 5.6 THINK AND DISCUSS

**Q5.16** Discuss what long-term solutions could be developed to be biologically and biomechanically active and reconstructive against alteration in physiological loading and biological conditions at the bone-implant interface.

**Q5.17** Discuss how osseoconductivity could be improved after couple of weeks, months, or even years of implantation.

## REFERENCES

[1] A.R. Gazdag, et al., Alternatives to autogenous bone graft: efficacy and indications, J. Am. Acad. Orthop. Surg. 3 (1) (1995) 1–8.
[2] Y. Liu, K. De Groot, E.B. Hunziker, BMP-2 liberated from biomimetic implant coatings induces and sustains direct ossification in an ectopic rat model, Bone 36 (5) (2005) 745–757.
[3] W. Wang, Y. Ouyang, C.K. Poh, Orthopaedic implant technology: biomaterials from past to future, Ann. Acad. Med. Singap. 40 (5) (2011) 237.
[4] E. Kon, et al., Platelet-rich plasma (PRP) to treat sports injuries: evidence to support its use, Knee Surg. Sports Traumatol. Arthrosc. 19 (4) (2011) 516–527.
[5] J. Alsousou, et al., The biology of platelet-rich plasma and its application in trauma and orthopaedic surgery, Bone Joint J. 91 (8) (2009) 987–996.
[6] B.J. Cole, et al., Platelet-rich plasma: where are we now and where are we going? Sports Health 2 (3) (2010) 203–210.
[7] R.E. Marx, et al., Platelet-rich plasma: growth factor enhancement for bone grafts, Oral Surg. Oral Med. Oral Pathol. Oral Radiol. Endod. 85 (6) (1998) 638–646.
[8] A. Roffi, et al., Does PRP enhance bone integration with grafts, graft substitutes, or implants? A systematic review, BMC Musculoskelet. Disord. 14 (1) (2013) 330.
[9] D. Chen, et al., Osseointegration of porous titanium implants with and without electrochemically deposited DCPD coating in an ovine model, J. Orthop. Surg. Res. 6 (1) (2011) 56.
[10] S.N. Papageorgiou, et al., Comparative effectiveness of natural and synthetic bone grafts in oral and maxillofacial surgery prior to insertion of dental implants: systematic review and network meta-analysis of parallel and cluster randomized controlled trials, J. Dent. 48 (2016) 1–8.

[11] S. Paderni, S. Terzi, L. Amendola, Major bone defect treatment with an osteoconductive bone substitute, Musculoskelet. Surg. 93 (2) (2009) 89–96.
[12] M. Szpalski, R. Gunzburg, Applications of calcium phosphate-based cancellous bone void fillers in trauma surgery, Orthopedics 25 (5) (2002) S601–S609.
[13] P. Wang, et al., Bone tissue engineering via nanostructured calcium phosphate biomaterials and stem cells, Bone Res. 2 (2014) 14017.
[14] A.R. Amini, C.T. Laurencin, S.P. Nukavarapu, Bone tissue engineering: recent advances and challenges, Crit. Rev. Biomed. Eng. 40 (5) (2012) 363–408.
[15] R.J. O'Keefe, J. Mao, Bone tissue engineering and regeneration: from discovery to the clinic—an overview, Tissue Eng. B Rev. 17 (6) (2011) 389–392.
[16] P. Ducheyne, R.L. Mauck, D.H. Smith, Biomaterials in the repair of sports injuries, Nat. Mater. 11 (8) (2012) 652–654.
[17] A.F. Steinert, et al., Concise review: the clinical application of mesenchymal stem cells for musculoskeletal regeneration: current status and perspectives, Stem Cells Transl. Med. 1 (3) (2012) 237–247.
[18] W.J. Landis, F.H. Silver, Mineral deposition in the extracellular matrices of vertebrate tissues: identification of possible apatite nucleation sites on type I collagen, Cells Tissues Organs 189 (1-4) (2009) 20–24.
[19] X. Su, et al., Organization of apatite crystals in human woven bone, Bone 32 (2) (2003) 150–162.
[20] D. Tadic, M. Epple, A thorough physicochemical characterisation of 14 calcium phosphate-based bone substitution materials in comparison to natural bone, Biomaterials 25 (6) (2004) 987–994.
[21] C.K. Chan, et al., Biomimetic nanocomposites for bone graft applications, Nanomedicine 1 (2) (2006) 177–188.
[22] M. Bohner, Design of ceramic-based cements and putties for bone graft substitution, Eur. Cell. Mater. 20 (1) (2010) 3–10.
[23] J. Zhang, et al., Calcium phosphate cements for bone substitution: chemistry, handling and mechanical properties, Acta Biomater. 10 (3) (2014) 1035–1049.
[24] M.-P. Ginebra, et al., New processing approaches in calcium phosphate cements and their applications in regenerative medicine, Acta Biomater. 6 (8) (2010) 2863–2873.
[25] W.E. Brown, A New Calcium Phosphate, Water-Setting Cement, Cements Research Progress, 1987.
[26] C.D. Friedman, et al., BoneSource™ hydroxyapatite cement: a novel biomaterial for craniofacial skeletal tissue engineering and reconstruction, J. Biomed. Mater. Res. A 43 (4) (1998) 428–432.
[27] H. Zhou, M.D. Weir, H.H. Xu, Effect of cell seeding density on proliferation and osteodifferentiation of umbilical cord stem cells on calcium phosphate cement-fiber scaffold, Tissue Eng. A 17 (21-22) (2011) 2603–2613.
[28] M.D. Weir, H.H. Xu, Osteoblastic induction on calcium phosphate cement–chitosan constructs for bone tissue engineering, J. Biomed. Mater. Res. A 94 (1) (2010) 223–233.
[29] M. Tang, et al., Human induced pluripotent stem cell-derived mesenchymal stem cell seeding on calcium phosphate scaffold for bone regeneration, Tissue Eng. A 20 (7-8) (2014) 1295–1305.
[30] M. Ramalingam, et al., Nanofiber scaffold gradients for interfacial tissue engineering, J. Biomater. Appl. 27 (6) (2013) 695–705.
[31] E. Martinez, et al., Effects of artificial micro-and nano-structured surfaces on cell behaviour, Ann. Anat. Anat. Anz. 191 (1) (2009) 126–135.
[32] M.S. Lord, M. Foss, F. Besenbacher, Influence of nanoscale surface topography on protein adsorption and cellular response, Nano Today 5 (1) (2010) 66–78.
[33] M.V. Jose, et al., Aligned bioactive multi-component nanofibrous nanocomposite scaffolds for bone tissue engineering, Macromol. Biosci. 10 (4) (2010) 433–444.

[34] S.-H. Kang, et al., Comparison of osseointegration between laser-etched and magnesium-incorporated oxidized implants in the rabbit femur, Int. J. Oral Maxillofac. Implants 28 (3) (2013) 775–781.
[35] Y.T. Sul, et al., Oxidized, bioactive implants are rapidly and strongly integrated in bone. Part 1–experimental implants, Clin. Oral Implants Res. 17 (5) (2006) 521–526.
[36] S. Galli, et al., Osteoconductive potential of mesoporous titania implant surfaces loaded with magnesium: an experimental study in the rabbit, Clin. Implant. Dent. Relat. Res. 17 (6) (2015) 1048–1059.
[37] N. Harmankaya, et al., Raloxifene and alendronate containing thin mesoporous titanium oxide films improve implant fixation to bone, Acta Biomater. 9 (6) (2013) 7064–7073.
[38] V. Bucci-Sabattini, et al., Effect of titanium implant surface nanoroughness and calcium phosphate low impregnation on bone cell activity in vitro, Oral Surg. Oral Med. Oral Pathol. Oral Radiol. Endod. 109 (2) (2010) 217–224.
[39] R. Jimbo, et al., Genetic responses to nanostructured calcium-phosphate-coated implants, J. Dent. Res. 90 (12) (2011) 1422–1427.
[40] A. Quaranta, et al., A histomorphometric study of nanothickness and plasma-sprayed calcium–phosphorous-coated implant surfaces in rabbit bone, J. Periodontol. 81 (4) (2010) 556–561.
[41] A. Wennerberg, et al., In vivo stability of hydroxyapatite nanoparticles coated on titanium implant surfaces, Int. J. Oral Maxillofac. Implants 26 (6) (2011) 1161–1166.
[42] Y.T. Sul, et al., The roles of surface chemistry and topography in the strength and rate of osseointegration of titanium implants in bone, J. Biomed. Mater. Res. A 89 (4) (2009) 942–950.
[43] Y.T. Sul, et al., Experimental evidence for interfacial biochemical bonding in osseointegrated titanium implants, Clin. Oral Implants Res. 24 (A100) (2013) 8–19.
[44] R. Jimbo, et al., Enhanced initial cell responses to chemically modified anodized titanium, Clin. Implant. Dent. Relat. Res. 10 (1) (2008) 55–61.
[45] Y.-T. Sul, et al., Optimum surface properties of oxidized implants for reinforcement of osseointeg ration: surface chemistry, oxide thickness, porosity, roughness, and crystal structure, Int. J. Oral Maxillofac. Implants 20 (3) (2005) 349–359.
[46] J.E. Ellingsen, et al., Improved retention and bone-to-implant contact with fluoride-modified titanium implants, Int. J. Oral Maxillofac. Implants 19 (5) (2004) 659–666.
[47] T. Sawase, et al., Application of oxygen ion implantation to titanium surfaces: effects on surface characteristics, corrosion resistance, and bone response, Clin. Implant. Dent. Relat. Res. 3 (4) (2001) 221–229.
[48] M. Leidi, et al., Nitric oxide mediates low magnesium inhibition of osteoblast-like cell proliferation, J. Nutr. Biochem. 23 (10) (2012) 1224–1229.
[49] E. Abed, R. Moreau, Importance of melastatin-like transient receptor potential 7 and cations (magnesium, calcium) in human osteoblast-like cell proliferation, Cell Prolif. 40 (6) (2007) 849–865.
[50] Y.-T. Sul, C. Johansson, T. Albrektsson, Which surface properties enhance bone response to implants? Comparison of oxidized magnesium, TiUnite, and Osseotite implant surfaces, Int. J. Prosthodont. 19 (4) (2006) 319–328.
[51] Y. Sul, et al., Bone tissue responses to Mg-incorporated oxidized implants and machine-turned implants in the rabbit femur, J. Appl. Biomater. Biomech. 3 (1) (2005) 18–28.
[52] L.R. Cho, et al., Bone response of Mg ion-implanted clinical implants with the plasma source ion implantation method, Clin. Oral Implants Res. 21 (8) (2010) 848–856.
[53] J.E. Davies, Understanding peri-implant endosseous healing, J. Dent. Educ. 67 (8) (2003) 932–949.
[54] J. Karlsson, et al., In vivo biomechanical stability of osseointegrating mesoporous TiO 2 implants, Acta Biomater. 8 (12) (2012) 4438–4446.

[55] K.-H. Kim, N. Ramaswamy, Electrochemical surface modification of titanium in dentistry, Dent. Mater. J. 28 (1) (2009) 20–36.
[56] H.-G. Simank, et al., The influence of surface coatings of dicalcium phosphate (DCPD) and growth and differentiation factor-5 (GDF-5) on the stability of titanium implants in vivo, Biomaterials 27 (21) (2006) 3988–3994.

# CHAPTER 6

# Osseoconductive Bioactive Surface Layering for Enhancement of Osseointegration

## 6.1 INTRODUCTION

The current orthopedic implants are fabricated from bioinert materials for long-term load-bearing against the physiological loading conditions. Majority of the load bearing orthopedic implants are developed from titanium alloy or polyetheretherketone (PEEK) materials. The titanium alloy has the potential of osseointegration enhancement much better than PEEK polymer. In order to enhance the bioactivity of the orthopedic implant at the contact interfaces with the bone tissue, the bioactive materials are coated on the surface of the implant to promote the bone in-growth and on-growth after implantation. In some surgical treatments of the bone injuries bone generation at the injured zone is targeted, while in other type of surgical treatments bone integration to the implant is a prominent clinical need. Various types of orthopedic treatment are given in Table 6.1 and the essential treatment for each type is highlighted. It could be revealed that the bone-implant integration was found to be a golden factor for effective treatment of the bone injuries to achieve early and long-term successful clinical outcomes. In fact, the quality of surgical treatments is being developed by incorporation of biological capabilities of the implant with biomechanical or mechanical advantages. It has been found that the pure consideration of mechanical stiffness or strength of the implant would not deliver comprehensive desired clinical outcomes and patients are not satisfied in their normal and special activities that they are dealing in their life, work, and entertainments [1, 2]. Although biological improvement of the orthopedic implants have been investigated and in some cases commercialized; however, there would be various types of limitations that affect the effectiveness of the utilized methods. One of such methods is the formation of a bioactive layer on the surface of the implant to establish an osseoconductive interface between the bone and implant. In this chapter, various types of bioactive

*Osseointegration of Orthopaedic Implants*
https://doi.org/10.1016/B978-0-12-813384-2.00006-0

**Table 6.1** Various types of severe bone injuries with associated implantation treatment and available implant or grafting system

| Injury type | Essential treatment | Implant or grafting system |
|---|---|---|
| Bone fracture with fitted gap | Bone generation at fracture gap | Trauma fixation |
| Bone fracture with unfitted gap | Bone generation at unfitted fracture gaps | Trauma fixation system<br>Bone grafting |
| Bone fracture with bone defect | Bone generation into bone substitute | Trauma fixation system<br>Bone substitute |
| Disc degeneration<br>Vertebral degeneration | Bone generation for vertebral fusion | Intervertebral cage<br>Vertebral replacement<br>Screw-rod fixation |
| Disc degeneration | Integration of disc prosthesis replacement to adjacent vertebrae | Disc replacement |
| Joint cartilage degeneration | Integration of joint replacement to the connective bones at the joint | Joint replacement |
| Bone defect due to tumor | Bone generation through the bone substitute | Bone substitute |
| Mild soft skeletal degeneration | Tissue generation through the tissue engineering scaffolds | Tissue engineering scaffold |

materials and methods are reviewed and their effectiveness to enhance the bioactivity of orthopedic implants is discussed. It is worth mentioning that the methods for porous surface layering of the orthopedic implants would be investigated as the other main category of the development concept which will be reviewed in Chapter 7.

## 6.2 CHALLENGES OF BIOACTIVE LAYERING ON THE IMPLANT SURFACE

Various parameters are involved in the evaluation of adhesion strength and bioactivity efficacy of the established bioactive layer on the surface of the implants. The bioactive layer is formed in microlevel, which would be looked as a coating layer on the implant surface. Implant with bioactive coating layer is commercially attractive compared to uncoated implants;

however, the extent of in vivo bioactivity of the implant might not be significant as expected or desired. The challenging or controversial aspects of bioactive layering on the surface of the implants has been observed as a poor adhesion of the coated layer to the implants, rapid adsorption of the coated layer in body fluid, temporary effect at early implantation, poor integration of the coated layer to the bone due to the presence of physiological loading conditions at the bone-implant interface, low porosity of coated layer matrix, inadequate or nonoptimal thickness of the coated layer, inhomogeneous-coated layer through the bone-implant contact surface, contamination of the coated layer during the coating process, and high brittleness of the coated layer (high brittleness of the coated layer would result in its cracking under shear or compression stress at the bone-implant interface). The description of these challenging aspects is provided in Table 6.2. To overcome these undesired outcomes, various types of layering methods with different process conditioning are being investigated. Although, bioactive layering of orthopedic implants is under development; however, in vitro and in vivo biological investigations revealed the better bioactivity of the processed implant compared to the unprocessed implants for various types of layering methods and materials. Indeed, this achievement is somehow obvious and would not be a significant outcome nowadays.

**Table 6.2** Challenging or controversy aspects of bioactive layering on the surface of the implants

| Aspect | Description |
|---|---|
| Poor adhesion of the coated layer to the implants | Poor adhesion of the coated layer to the implant would cause release of the coated layer at the bone-implant interface. If the released material would be bioactive, the long-term effectiveness of the coated layer is affected. From which, the bone generation is reduced and osseointegration might not be achieved effectively. |
| Rapid adsorption of the coated layer in body fluid | Rapid adsorption of the coated layer would reduce the long-term effect of the bone remodeling process at the bone-implant interface. Even this may reduce the effectiveness conduction of the osteogenic cell to |

*Continued*

**Table 6.2** Challenging or controversy aspects of bioactive layering on the surface of the implants—cont'd

| Aspect | Description |
|---|---|
| | be adhered and proliferated on the implant surface at an early implantation. |
| Temporary effect at short-term implantation | The bioactive layering on the surface of the implant would have short-term osseoconductivity effects in the formation of the bone tissue on the surface of the implant. Therefore, solutions such as creation of a bioinert or long-degradation host layer to coat the bioactive materials is encouraged. Likewise, nano-thickness coating of the nanoparticles of the bioactive material is the other solution that would be promising to enhance the long-term effectiveness of the osseointegration. In this chapter and Chapter 7 these solutions are described. |
| Poor integration of the coated layer to the bone due to the presence of physiological loading conditions at the bone-implant interface | Physiological loading condition (particularly shear stress) at the bone-implant interface may cause weakening of the osseointegration in cases with poor biological and mechanical integration between the bone and the implant. |
| Low porosity of the coated layer matrix | Low porosity of the coated layer would reduce the flow of blood through the porous surface and thus reduce the effectiveness of the formed porous layer in conduction of the bone remodeling on the surface of the implant. |
| Inadequate or non-optimum thickness of the coated layer | Based on the implantation treatment type or according to the method of the implantation or replacement of the implant, the thickness of the coating layer might be different. The thickness greater than 100 μm is optimum for total joint replacement while thickness less than 100 μm would be beneficial for insertion of screw-type implants. Likewise, |

**Table 6.2** Challenging or controversy aspects of bioactive layering on the surface of the implants—cont'd

| Aspect | Description |
|---|---|
| | thickness less than 1 μm has advantages for implantations with impaction force such as insertion of the intervertebral fusion cage between the vertebral bodies. |
| Inhomogeneous-coated layer through the bone-implant contact surface | Inhomogeneous-coated layer would cause deterioration of the osseointegration at some areas and thus increase the risk of osseointegration failure at the well-coated areas. |
| Contamination of the coated layer during the coating process | Contamination of the coated layer is greatly destructive in successful osseointegration of the implant. Thus, care must be taken to process the coating layer in an isolated area without bacterial microorganisms and other contaminants that might be seated on the coating surface. |
| High brittleness of the coated layer | High brittleness of the coated layer would result to its cracking under shear or compression stress at the bone-implant interface |

## 6.3 CALCIUM PHOSPHATES AS THE MAIN COMMERCIAL OSSEOCONDUCTIVE COATING LAYER

Due to poor surface energy of the orthopedic implants (particularly, hydrophobic materials such as PEEK), adhesion of osteogenic cells, proteins, growth factors, and other biological factors would not be achieved effectively [3–9]. Specifically, this would result in poor adhesion of fibronectin protein and therefore adhesion and osseoconduction of osteoblast cells for the generation of bone tissue on the surface of the implant is not adequately governed. In order to increase the surface energy and hydrophilic characteristics of the orthopedic implants, coating of calcium phosphate bioceramics [e.g., HA, bioglass, tricalcium phosphate (TCP)] or osseoconductive metallic materials (e.g., titanium, tantalum, magnesium) have been developed (the latter is reviewed in Chapter 7). The calcium phosphates are categorized as bioceramics and synthesized in various formulations. Since higher portion of the bone inorganic constitute is structured from calcium phosphate, adsorption of bone cellular matrix to the bioceramic-coated layer is induced. In addition, due to inherent brittleness of the ceramic-based materials, the presence of huge number of nano-boundaries at the surface of synthesized calcium phosphate-coated layer would raise the surface energy of the implant at the coated area. Furthermore, the processing of a thin layer in nano or microthickness on the substrate would result in the formation of amorphous-crystalline structure with pores (the porosity of the coated layer is an advantage and in some coating processes, porous coating is targeted for the enhancement of bone in-growth).

Calcium phosphates are classified based on the ratio of the calcium over phosphor elements (Ca/P ratio). This ratio could be altered from below the ratio of 1.00 to over the ratio of 2.00. Table 6.3 would show some calcium phosphate materials with different formulation and Ca/P ratio. As given in this table, each calcium phosphate material has different solubility rate in human body fluid. The formulation of $Ca_{10}(PO_4)_6(OH)_2$ which is known as the hydroxyapatite (HA), has the lowest rate of solubility which would increase the longevity of the HA degradation compared to other calcium phosphates. This might be the reason that ISO 13779 would recommend the HA with ratio of the 1.67 as the optimum Ca/P ratio for coating of the calcium phosphate on the surface of the implant. In general, formation of the calcium phosphate with unique Ca/P ratio might not be achievable through the processing methods such as plasma spraying or plasma electrolyte oxidation (PEO). The current major coating process is plasma spraying process which is now commercially utilized for the formation of HA coating layer on the surface of the implant. Although, the orthopedic implant manufacturers would extensively utilize this method, however, not all challenging aspects of the coating layer (as highlighted in Table 6.2) in early and long-term osseoconductivity might be obtainable via this method. Through this method, the irregular formation of calcium phosphate formulations with different ratio of Ca/P has been reported. Heimann et al. [10] measured the Ca/P ratio in two sections of the HA coating layer using electron probe microanalyzer (EPMA). It was observed that the Ca/P ratio is altered from 1.00 to upper than 2.00 which would show the inhomogeneous formation of the calcium phosphate through the coating layer (Fig. 6.1, left). With controlling and optimizing of the plasma spraying process parameters, the average of the Ca/P ratio could be achieved 1.67 to be closed as that of recommended in ISO 13779. The other disadvantage of the plasma

**Table 6.3** Calcium phosphate materials with different formulation, Ca/P ratio, and solubility

| Ca/P molar ratio | Chemical formula | Typical abbreviations | Solubility at 25°C, log Ks |
|---|---|---|---|
| 0.5 | $Ca(H_2PO_4)_2.H_2O$ | MCPM | 1.14 |
| 0.5 | $Ca(H_2PO_4)_2$ | MCPA or MCP | 1.14 |
| 1.0 | $CaHPO_4.2H_2O$ | DCPD | 6.59 |
| 1.0 | $CaHPO_4$ | DCPA or DCP | 6.90 |
| 1.33 | $Ca_8(HPO_4)_2(PO_4)_4.5H_2O$ | OCP | 96.6 |
| 1.5 | $\alpha$-$Ca_3(PO_4)_2$ | α-TCP | 25.5 |
| 1.5 | $\beta$-$Ca_3(PO_4)_2$ | β-TCP | 28.9 |
| 1.2–2.2 | $Ca_xH_y(PO_4)_z.nH_2O$ $n=3$–$4.5$; 15%–20% $H_2O$ | ACP | |
| 1.5–1.67 | $Ca_{10-x}(HPO_4)_x(PO_4)_{6-x}(OH)_{2-x}$ $(0<x<1)$ | CDHA or Ca-def HA | ~85 |
| 1.67 | $Ca_{10}(PO_4)_6(OH)_2$ | HA, Hap or OHAp | 116.8 |
| 1.67 | $Ca_{10}(PO_4)_6F_2$ | FA or FAp | 120 |
| 1.67 | $Ca_{10}(PO_4)_6O$ | OA, OAp, or OXA | ~69 |
| 2.0 | $Ca_4(PO_4)_2O$ | TetCP | 38–44 |

(Courtesy of M. Canillas, P. Pena, A.H. de Aza, M.A. Rodríguez, Calcium phosphates for biomedical applications, Bol. Soc. Esp. Ceram. Vidrio 56 (2017) 91–112 and permission from Elsevier Publisher.)

sprayed-coated layer is the poor adhesion of the generated new bone tissue at the bone-implant interface in long-term implantation. In fact, the HA constitute is degraded in human body fluid to conduct formation of the new bone tissue through the implant surface. When the HA-coated layer is entirely degraded, the generated bone tissue would be in direct contact with the implant surface. In case of the smooth surface topology of the implant, the integration of the bone tissue and the implant surface might not be as strong as when the coated layer has been deposited to the implant surface.

Among various types of processing methods for formation of the HA, PEO has been recently under interest to form the bilayer of $TiO_2$ and HA on the surface of the implant [10–13]. This technology was first developed for the dental implant to reduce the period of effective osseointegration at the interface of the mandibular bone and the dental implant. However, such coating process was also utilized for bioactivity enhancement of the pedicle screw (Fig. 6.2) and could be developed for the creation of HA

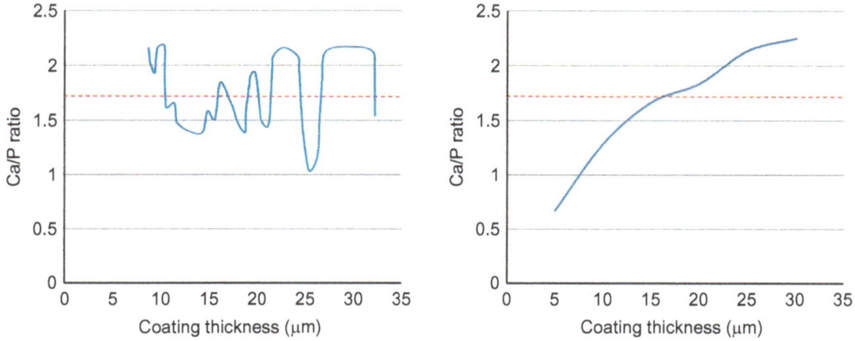


**Fig. 6.1** Schematic illustration of the Ca/P ratio in the line of coating thickness which is processed by (left) plasma spraying process, and (right) plasma electrolyte oxidation (PEO).

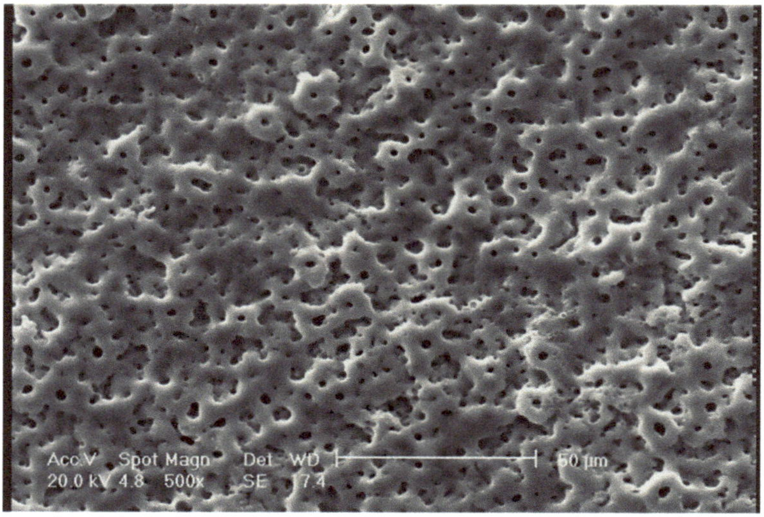


**Fig. 6.2** Scan electron microscopy of the HA-$TiO_2$ layer on the surface of the titanium substrate.

coating layer on the surface of the cementless components in total joint replacements. This process would be developed on some materials such as titanium, aluminium, and magnesium that could be anodized by oxidation of the surface layer. In this process, the surface of the substrate is first anodized to construct a porous structure to establish a host layer for the bone in-growth (Fig. 6.2). For instance, this porous construct for titanium substrate would be titanium dioxide ($TiO_2$). Then the HA is constituted through the porous layer and could be continued until the complete filling of the created pores of the oxide layer. The main advantage of the established

$TiO_2$-HA layer by PEO method compared to plasma sprayed HA-coated layer is that the generated new bone tissue could be penetrated through the pores of the oxide layer in long-term implantation and thus increase the adhesion strength between the bone and the implant to establish more rapid and effective osseointegration of the implant (establishment of the porous Ti or $TiO_2$ layer on the surface of the implant is reviewed as osseoconductive porous surface layering in Chapter 7). The other advantage might be the formation of homogeneous calcium phosphate constitutes through the thickness of the coated layer. This achievement was found in the investigation carried out by Engineer Naderi (Tajhiz Sanat Rahnama Co., Isfahan, Iran) when he was developing PEO process for the creation of $TiO_2$-HA constitute on the titanium pedicle screw. The surface of the processed implant was analyzed by EDAX (energy dispersive X-ray analysis) method to find out the Ca/P ratio for various thickness values. Interestingly, it was observed that for lower thickness of the $TiO_2$-HA (about 1–5 μm), the Ca/P ratio is in the range of 0.6–1 and it would be increased monotonically by increasing the $TiO_2$-HA thickness up to Ca/P ratio of 2.26 for thickness about 30–35 μm (Fig. 6.1, right). The trend of Ca/P ratio in relation with the thickness of the $TiO_2$-HA layer is illustrated in Fig. 6.1 (the Naderi's outcomes have not been published yet and no similar published results could be found in the literature to reveal his findings, thus it should be considered as a hypothesis for further clarification).

## 6.4 LOADING OF PROTEIN-BASED MATERIALS ON THE IMPLANT SURFACE

The immobilization or loading of protein-based materials has been found effective in the enhancement of osseointegration [14–19]. Particularly, bone morphogenetic protein-2 and -7 (BMP-2 and BMP-7) have been addressed as the proteins for induction of osteoblast cells activity [18, 20]. These proteins would be compounded with synthesized calcium phosphates as bone graft for induction of bone formation in the treatment of spinal fusion or bone defect [21–23]. The BMPs might be loaded through the surface of the implant. In order to load the BMPs, a host layer on the implant surface is needed for absorption of the BMPs. Various materials such as collagen gels, chitosan, etc. have been addressed in the literature as the host layer for BMP loading on the implant surface [18, 24, 25]. The host layer absorbs the BMPs at in-vitro environment or preimplantation and releases the BMPs at the interface of bone-implant postimplantation. The key factor of host layer is to control the release of BMP in correlation with effective rate of bone formation at the bone-implant interface. The rapid release of BMP would not have bone formation advantageous on the surface of the implant. Therefore, binding of the BMP to the host layer for longer time is desired for loading of BMP on the implant surface. In this respect, heparin was addressed as an effective alternative to be used as host layer as it has great affinity to bind to vascular, fibroblast, and transferring growth factors (known as VEGF, bFGF, and TGF-β) [26]. As highlighted in the literature, heparin has a great control advantages for binding to the BMPs and therefore prolongs the release of BMPs [27, 28]. The process of loading or deposition of heparin has been described for titanium implant by Lee et al. [18]. Based on their report, the implant is immersed in Hep-DOPA solution and slowly shaken in darkness to achieve heparin-treated Ti (for further details, it is recommended to review the study of Lee et al.). After processing of heparin-Ti, BMP would be immobilized on the surface by immersion of heparin-Ti implant to the 0.1 M MES buffer and then BMP protein is added to the buffer and then letting for binding of BMP to the created heparin matrix on the implant surface (Fig. 6.3). In order to evaluate the osteoblast activity of the BMP-enhanced implant, proliferation of osteoblast cells, activity of ALP, and deposition of calcium at the implanted zone would be investigated. In addition to protein immobilization on the implant surface, drug delivery advantage of coating layer on the implant surface has been studied in recent years [29]. The presence of antibacterial factor on the coating layer would be one of the drug delivery strategies to improve osseointegration of implant to the bone where the risk of infection at the operated zone is considerable. The antibacterial adhesion of orthopedic implant could be enhanced by immobilization of antibacterial drugs (e.g., gentamicin sulfate). The micro-organism bacterial agents might be adhered to the implant surface intraoperatively or even postoperatively from the sutured incision at the operated zone. The growth of bacteria on the implant surface would cause creation of a thick biofilm which may prevent host immune system or disaffect penetration of antibiotics that is administered postoperatively [18]. It is beneficial to compound the drug particles through the bioactive degradable material (e.g., HA, collagen, or chitosan) to prevent rapid drug delivery at the interface of the bone and the implant and to keep the effect of drug treatment for longer period after the implantation. Fig. 6.4 displayed the schematic concept of drug delivery where the drug particles are compounded with a bioactive degradable material.

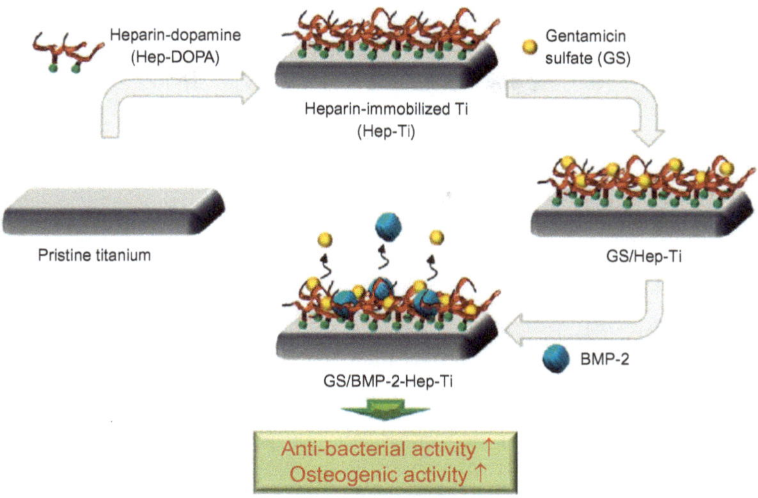


**Fig. 6.3** Process flow of GS/BMP-2-heparin-Ti layering on the surface of titanium substrate for enhancement of osseoconductivity. *(Courtesy of D.-W. Lee, et al., Gentamicin and bone morphogenic protein-2 (BMP-2)-delivering heparinized-titanium implant with enhanced antibacterial activity and osteointegration, Bone 50(4) (2012) 974–982 and permission from Elsevier Publisher.)*

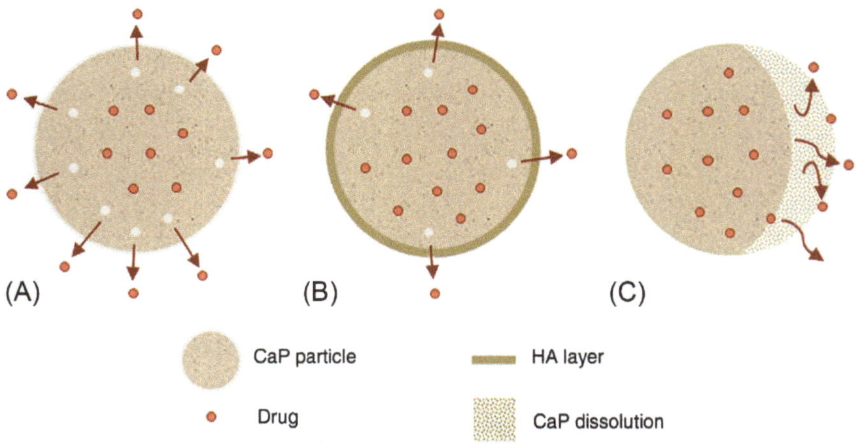


**Fig. 6.4** Schematic concept of drug delivery where the drug particles are compounded with a bioactive degradable material (A–C). *(Courtesy of M. Canillas, P. Pena, A.H. de Aza, M.A. Rodríguez, Calcium phosphates for biomedical applications, Bol. Soc. Esp. Ceram. Vidrio 56 (2017) 91–112 and permission from Elsevier Publisher.)*

## 6.5 HA/COLLAGEN NANOCOMPOSITE AS AN OSSEOCONDUCTIVE LAYER ON IMPLANT SURFACE

Uezono et al. [30] investigated the superior osseointegration ability of the HA/collagen nanocomposite which had been coated on the titanium substrate. As reviewed in Section 5.6, this nano-composition is similar to the bone composition at which collagen fibril is the organic and HA is the inorganic portions of the lamellar bone. They showed that formation of fibrous tissue at the bone-implant interface would be significantly reduced compared to the HA-coated and -uncoated titanium implants. Likewise, the extent of new bone formation is much higher than that of HA-coated implant. As they have illustrated, no fibrous tissue is observed in greater area of bone-implant contact surface in HA/collagen-coated implants compared to HA-coated and -uncoated implants (it is recommended to see the Fig. 5 of the Uezano et al.'s study [30], which shows the osseointegration of the titanium implants; uncoated, HA coated, and coated with nanocomposite of HA/collagen).

Furthermore, a significant bonding strength of the implant to the bone was revealed for HA/collagen-coated implant through the shear testing experiments (for further details, this reference [30] is highly recommended to study). As demonstrated in Fig. 8 Ref. [30], due to the strong attachment of the HA-/collagen-coated implant to the high dense bone tissue, removal of the implant is achieved in higher force compared to HA-coated implant. According to the removal test carried out by Uezano et al., the bone matrix at the interface is fractured by removal of the HA-/collagen-coated implant from the bone and larger segment of the bone is detached compared to HA-coated implant which would show the higher strength of the integrated bone to the implant surface. One of the advantages of HA/collagen is to develop the on-plant device that is inserted under the periosteum on one side and on the bone on the other side such as the implants that are placed on cranio-maxillofacial bones. From what has been observed in Uezono's et al. Study, the enhanced osseoconductive advantages of mixed HA and collagen coating layer could be concluded for on-plant devices. Coating of HA/collagen was processed via dip coating method at which the Ti substrate is immersed in the HA/collagen suspension followed by dehydrating by ethanol and allowed for cross-linking process in oven at the temperature of 140°C for 12h. Due to the high shear strength of the HA/collagen coating layering, this process could be found as a promising method for the development of high osseointegrant implant. The interesting point of this method is the presence of collagen fibers that could prolong resorption of nano-HA particles at the bone-implant interface in long-term implantation.

## 6.6 CHITOSAN AS A GREAT OSSEOCONDUCTIVE SURFACE LAYERING

Coating of resourced materials from animals has been one of the development concepts for use in the enhancement of bioactivity of orthopedic implants. Chitosan is one of these materials that have been introduced for promotion of bone-implant integration as a bioactive material [31–34]. Chitosan is the derivative of chitin which is found in external covering tissue of arthropods' body. It is largely existed in the shells of the sea animals such as lobsters, shrimp, and crabs. After removing the calcium-based minerals and proteinaceous with acid and alkali, respectively, the remaining constitute or composition is treated with stronger alkali to remove more than 50% of the acetyl through which the residual composition that is obtained is known as chitosan. The molecular structure of the chitosan is similar to polymeric materials due to its chain-based structuring [35]. The percentage of acetyl removal could be organized up to 100%. The process of acetyl extraction of chitin is called deacetylation [the level of deacetylation is termed as degree of deacetylation (DDA)]. Chitosan has similar chemical structuring with effective biological factors in conduction of the bone remodeling process (e.g., hydluronic acid and glycosaminoglycan molecules [31]). As highlighted in the literature [36, 37], such chemical structure could play as a functional reactor of the collagen fibres organization during reconstruction of the bone tissue. Therefore, chitosan has biological osseoinductive effect during transmission of osteoblast cell to the osteocyte cell at which collagen fibers are extended from the osteoblast cell to its surrounding to create a mineralization platform for the formation of mature bone.

Chitosan has positive charge naturally, which increases the absorption of growth factors and cytokines with negative charge. This would be the advantages of chitosan osseoconductivity to increase the density of growth factors at the bone-implant interface [38–40]. The interesting factor for enhancement of this ability is to increase the level of acetyle extraction during synthesizing of the chitosan from which the positive charge of deacetylated chitosan is enhanced. This would conduct higher amount of growth factor attachment to the chitosan-coated implant, thus more differentiation and proliferation of osteoblast cells is resulted [41, 42]. Degradation of chitosan was also found to be effective in the enhancement of osseoinductivity by inducing the expression of the osteogenic genes [e.g., alkaline phosphatise and morphogenetic protein (MPB)] [43]. Degradation of chitosan has inverse proportion with the degree of deacetylation (DDA) [44]. This would be another advantage point of chitosan to be controllable in degradation and could be considered as a delivery system for the release of therapeutic agents (e.g., antibacterial drugs, growth factors, osteogenesis proteins, etc.). In fact, lower level of acetyle would enhance the molecular bonding of the polymeric phase in chitosan constitute, thus cause stronger polymeric chain pack with greater crystalline phase from which the rate of chitosan degradation is deducted. Overall, the remarkable advantages of chitosan would categorize it as a bioactive material with great osseoconductive and osseoinductive functionality from promoting the growth factors attachment and expression of osteogenic or osteoblastic genes to its controllable rate of degradation for drug delivery systems [45]. Therefore, the chitosan has been an attractive promising bioactive material for coating on the surface of the bioinert orthopedic implant and for development of tissue scaffolds, bone substitutes, or synthesized bone graft constitutes.

## 6.7 CHALLENGES OF THICK HA COATING ON LONG-STANDING OF OSSEOINTEGRATION

Plasma spraying of HA has good early effect to accelerate the primary stage of implant osseointegration [46]. It was found that in unfitted bone-implant interface with considerable gap, processed thick-coated layer of HA (100–400 μm) by plasma spraying method could enhance the osseoinduction and the osseoconduction to compensate the existing gap and ultimately the formation of bone at the bone-implant interface [47]. However, according to clinical observation, long-term failure of osseointegration might be caused due to the detachment of the coating layer from the implant [48, 49]. Detachment of thick HA layer may cause inflammation and resorption of bone at the interface and therefore, cause weakening of the created bone tissue on the surface of the implant. In fact, a thick layer of coating is achieved by several passage of plasma gun from which the coated layer is formed layer by layer on the implant surface. This would increase the creation of shear stress through the layers and induce initiation and propagation of crack under the long-term dynamic physiological loading conditions which is transferred through the bone-implant interface. From the histological observations by Registal et al. [49], adhesion of the bone osteogenic cells and the formed extracellular bone matrix was detected to the detached HA particles from the implant surface. This would deteriorate the process of osseointegration, which reveals the undesired effect of thick HA coating layer and reduction of biomechanical strength in long-term implantation, which was observed by Registal et al. [49] after 52 weeks of implantation in a rabbit tibia. In order to overcome the long-term disadvantage of thick plasma-sprayed coating layer, nanosized HA particles or crystals have been utilized to form a mono-coating layer on the surface of the implants, particularly, hydrophobic materials such as PEEK polymer [50–52]. In fact, due to the low level of surface energy at the surface of hydrophobic materials, bonding strength of the coated layer and substrate would be much lower than the shear strength between the sublayers of the thick HA-coated layer. Therefore, the thick-coated layer might be detached or debonded from the hydrophobic surface in the long-term existence of dynamic physiological shear stress at the bone-implant interface.

## 6.8 SUPERIOR OSSEOINTEGRATION OF NANO-MONOLAYER OF HA AS A COATING LAYER

The enhanced solution to improve the long-term deposition of the HA-coated layer on PEEK implant is focusing on the development of its surface energy by retaining of the microroughness of the surface at an optimum level but effective for long-term implantation. This superior advantage would be achievable with the formation of nano-thickness of the nanocrystal HA in the range of 10–20 nm which has been reported recently [53, 54]. In this method of coating, a nano-thin monolayer was formed on the surface of the PEEK implant. Although, the amount of HA particles in manolayer of nano-HA is much lower than that of multilayer of micro-HA; however, higher range of bone to implant contact area was observed for nano-monolayer than micro-multilayer [55]. A long-term implantation study would be beneficial to show the superior effect of monolayer HA on the implant. The utilized method for formation of monolayer nano-HA is to apply 50-mL coating solution through the implant followed by rotation of implant at 2700 rpm for duration of 5 s. The coating solution is consisted of nano-HA crystals which would be suspended by the surfactants and the solvent. The length and width of the HA crystals are in the range of 20–50 nm and 2–10 nm. After completion of coating process, the implant is heated in an oxygen-enriched oven at 325°C for 5 min. Heat treatment with these conditions would facilitate removal of the surfactants and better deposition of HA crystals to the implant surface. This achievement would generate the idea of forming of the osseoconductive bilayer on the implant surface at which the first layer is a thick bioactive layer for the enhancement of early osseointegration at the contact interface of the bone and orthopedic implants. This is because bone-implant interface in implantation of orthopedic implants might not be fitted and there would be a gap at the interface which necessitates coating of thick bioactive layer. The second layer or underneath layer would be remained for long-term after detachment of the first or shell layer and withstand its stability under the physiological shear stress. The monolayer coating in nanoscale should be greatly beneficial in implantation of the dental implants, intervertebral PEEK fusion cages, and soft tissue implants while bilayer coating could enhance the long-term effectiveness of the osseointegration in implantation of the total joint replacements.

## 6.9 SUMMARY

One of the main methods to increase the osseoconductivity of the orthopedic implants is coating, immobilization, deposition, or loading of the bioactive materials (e.g., calcium phosphates, collagen, chitosan, magnesium, etc.) or biological factors (e.g., growth factors, protein, osteoblastic genes, etc.) that can conduct adherence of the osteogenic cells on the implant surface followed by proliferation of the osteoblast cells. Currently, coating of HA on the surface of various types of orthopedic implants (e.g., femoral stem, tibial tray, acetabular cup, pedicle screw, intervertebral fusion cage, etc.) is commercially extensive for the enhancement of the implant bioactivity at the contact surface with the bone tissue. As highlighted in Table 6.2, effectiveness of the osseoconductive layering would be associated with various types of parameters or challenges which would influence on longevity of the implant osseointegration. Challenges such as adhesion strength of the osseoconductive layer to the substrate and long-term effectiveness of the osseoconductivity could be highlighted as two examples of nine critical aspects that might be contemplated in the enhancement of implant osseointegration (Table 6.2). As reviewed in Section 6.3, various types of calcium phosphates might be synthesized (Table 6.3) with different formulations and Ca/P ratio and solubility rate in human body fluid. The solubility rate for Ca/P ratio of 1.67 is the minimal rate that would be recommended by ISO 13779 as the required Ca/P ratio for coating of HA on the surface of the implant. However, the ratio of 1.5 for the TCP has been recently found more effective in some cases. Plasma spraying of HA would be the major processing method that is currently commercialized by the orthopedic manufacturers. Due to disadvantages of the plasma spraying-coated layer, PEO method was utilized for coating of HA on materials with biocompatible oxide layer (e.g., titanium, magnesium, and aluminum). The PEO would establish a porous oxide layer of the substrate on the surface to enhance deposition of the bioactive material (e.g., HA, magnesium, chitosan, growth factors, etc.). The formed osseoconductive layer by PEO method would have advantages compared to plasma spraying method (refer to Section 6.3). Adhesion strength of the coated layer during implantation (particularly, purchasing of the screw through the bone) and effective integration of the mature or lamellar bone tissue to the porous surface of implant are two major advantages.

In addition to the current commercial processing of HA on the surface of the implant, the formation of other bioactive materials such as proteins, collagen, and chitosan have been described in this chapter. All these materials would enhance the bone remodeling process on the surface of the implant, thus creating bone tissue with higher mineral density. The other reviewed aspects were coating of nano-HA particles and formation of nano-thickness HA on the implant surface. In general, coating in nanoscale has been revealed to be more effective in adhesion of the osteogenic cells followed by proliferation and differentiation on the processed area. In general, an effective and successful formation of the osseoconductive layer on the surface of the orthopedic implants is beneficial to provide following five advantages:

**(1)** processing of coated material on the implant with good bonding strength (the coated material should not be removed or detached during implantation at the target-injured area),

**(2)** implantation of the implant with preservation of coated material with no or minimal reduction of coating thickness or bonding strength of coated layer (the coated material is essential to facilitate early bone formation effect to prevent early migration or dislocation of the implant in contact with bone),

**(3)** early effective osseoconductivity of the coating layer to enhance early integration or depositing of new lamellar bone formation on the implant surface,

**(4)** mild long-term enhanced osseoconductivity to preserve the biological integration of the implant during the treatment period (depending on the treatment type, this period might be between 4 and 12 months), and

**(5)** high long-term enhanced osseoconductivity to preserve the biological integration and reorganize the failed or weakened osseointegrated areas with the internal or external reactivation of osseoconductivity advantages.

Although, the characteristics of the host bone are not consistent or the severity and type of injury might not be similar; however, consideration of these five advantages is crucial to be contemplated in the development of new methods for the enhancement of implant bioactivity in contact surfaces with the bone tissue.

## 6.10 REMIND AND LEARN

**Q6.1** Discuss why promotion of the implant bioactivity would be crucial for early and long-term effectiveness of the implantation treatment.

**Q6.2** What types of severe bone injuries might be occurred? What is the essential implantation treatment for each injury type? What implant system could be utilized for each implantation treatment?

**Q6.3** What are the challenging or controversial aspects of bioactive layering on the surface of the implants? Describe each aspect.

**Q6.4** Discuss why the surface of orthopedic implants would not have the ability of the biological integration with the bone.

**Q6.5** What types of calcium phosphate might be formulated? What parameters would be normally used to distinguish the calcium phosphate formulations? How would calcium phosphate increase the surface energy of the implants?

**Q6.6** Why HA has been recommended by ISO 13779 for coating on the surface of the implant? What commercial processing methods might be utilized for coating of the HA on the surface of the orthopedic implant?

**Q6.7** What is the Ca/P ratio of the formed HA coating layer by plasma spraying and PEO coating methods? Discuss the advantages of Ca/P ratio in PEO method compared to plasma spraying method.

**Q6.8** What are the disadvantages of the plasma spraying coating in the long-term osseointegration of the implants? How would the effects of such disadvantages be reduced by PEO method?

**Q6.9** What are the characteristics of the processed PEO osseoconductive layer on the implant surface? What type of materials could be processed by PEO and why?

**Q6.10** What is the main type of protein that might be loaded or immobilized on the implant surface? What is the biological effect of this protein in enhancement of osseoconductivity at the bone-implant interface?

**Q6.11** How would the osseoinductive protein be loaded on the surface of the implant? Discuss the effectiveness of the loaded protein in enhancement of the osseointegration (express what types of materials might be used as the host layer for promotion of the protein osseoconductivity effectiveness).

**Q6.12** What are the advantages of the heparin to be used as the host layer for loading of the protein and growth factors on the surface of the implant? Describe the deposition or loading process of the heparin on the titanium implant.

**Q6.13** Express the benefits of antibacterial immobilization on the implant surface. What type of drug delivery could be beneficial to be organized on the surface of the orthopedic implants?

**Q6.14** How could osseointegration ability of the implant be superior when HA/collagen nanocomposite is used as coating layer? What is the effect of this coating material in formation of fibrous tissue at the bone-implant interface (consider study of the Fig. 5 in the study of Uezano et al. [30])?

**Q6.15** How could the bonding strength of the implant be evaluated? Compare the bonding strength of the implant with coating layers of HA/collagen and HA (see Fig. 8 of Ref.[30] for better observation of the implant after detachment from the bone).

**Q6.16** What type of implant would be greatly beneficial to be developed by coating of the HA/collagen? How could HA/collagen nanocomposite be processed on the titanium implant surface?

**Q6.17** Express the resource of chitosan in the nature. How chitosan would be synthesized from the natural resources? What is deacetylation in synthesizing of the chitosan?

**Q6.18** Discuss how chitosan could enhance the osseoconductivity of the implant surface. How would the attachment of the growth factors be increased by the existence of chitosan on the surface of the implant?

**Q6.19** What is the effect of acetyl in degradation of the chitosan in human body fluid?

**Q6.20** In what implantation treatment, thick layering of HA might be required and why? What are the disadvantages of the thick HA-coated layer? What are the advantages of thin monolayer of HA in nanoscale, particularly in hydrophobic surface?

**Q6.21** How the nano-monolayer of HA could be processed on the surface of the PEEK polymer?

**Q6.22** What is the concept of osseoconductive bilayer in bioactivity enhancement of the implants with thick HA-coated layer? In what types of implants the effective long-term osseointegration could be enhanced by coating of nano-monolayer from bioactive material? How about coating of osseoconductive bilayer?

## 6.11 THINK AND DISCUSS

**Q6.23** Discuss the current effectiveness of the surface bioactivity in formation of the bioactive coating layer on the surface of the orthopedic implants.

**Q6.24** Discuss how the osseointegration of the bioactive-coated implant could be weakened in long-term implantation.

## REFERENCES

[1] Y. Ren, et al., Microwave assisted coating of bioactive amorphous magnesium phosphate (AMP) on polyetheretherketone (PEEK), Mater. Sci. Eng. C 85 (2018) 107–113.
[2] H. Alghamdi, et al., Calcium-phosphate-coated oral implants promote osseointegration in osteoporosis, J. Dent. Res. 92 (11) (2013) 982–988.
[3] P. Johansson, et al., Biomechanical evaluation and surface characterization of a nano-modified surface on PEEK implants: a study in the rabbit tibia, Int. J. Nanomedicine 9 (2014) 3903.
[4] D.M. Devine, et al., Coating of carbon fiber-reinforced polyetheretherketone implants with titanium to improve bone apposition, J. Biomed. Mater. Res. B Appl. Biomater. 101 (4) (2013) 591–598.
[5] T.A. Kumar, J.B. Jei, B. Muthukumar, Comparison of osteogenic potential of poly-ether-ether-ketone with titanium-coated poly-ether-ether-ketone and titanium-blended poly-ether-ether-ketone: an in vitro study, J. Indian Prosthodont. Soc. 17 (2) (2017) 167.
[6] H.-K. Tsou, et al., In vivo osseointegration performance of titanium dioxide coating modified polyetheretherketone using arc ion plating for spinal implant application, Biomed. Res. Int. 2015 (2015) 1–9.
[7] P. Johansson, et al., Nanosized hydroxyapatite coating on PEEK implants enhances early bone formation: a histological and three-dimensional investigation in rabbit bone, Materials 8 (7) (2015) 3815–3830.
[8] W.R. Walsh, et al., Plasma-sprayed titanium coating to polyetheretherketone improves the bone-implant interface, Spine J. 15 (5) (2015) 1041–1049.
[9] S. Stübinger, et al., Titanium and hydroxyapatite coating of polyetheretherketone and carbon fiber-reinforced polyetheretherketone: a pilot study in sheep, J. Biomed. Mater. Res. B Appl. Biomater. 104 (6) (2016) 1182–1191.
[10] H.D. Lehmann, R.B. Heimann, Bioceramic Coating for Mecial Implants, Wiley-VCH Verlag GmbH & Co, 2015.
[11] M. Canillas, P. Pena, A.H. de Aza, M.A. Rodríguez, Calcium phosphates for biomedical applications, Bol. Soc. Esp. Ceram. Vidrio **56** (2017) 91–112.
[12] H.-.K. Tsou, P.-.Y. Hsieh, Anticorrosive, antimicrobial, and bioactive titanium dioxide coating for surface-modified purpose on biomedical material, in: Application of Titanium Dioxide, InTech, 2017, pp. 103–122.

[13] G.C.R. Wing Kiu Yeung, A. Matthews, A. Yerokhin, In vitro biological response of plasma electrolytically oxidized and plasma-sprayed hydroxyapatite coatings on Ti–6Al–4V alloy, J Biomed Mater Res B **101B** (6) (2013) 939–949.
[14] G. King, et al., Recombinant human bone morphogenetic protein-2 promotes wound healing in rat periodontal fenestration defects, J. Dent. Res. 76 (8) (1997) 1460–1470.
[15] G. Zellin, A. Linde, Importance of delivery systems for growth-stimulatory factors in combination with osteopromotive membranes. An experimental study using rhBMP-2 in rat mandibular defects, J. Biomed. Mater. Res. A 35 (2) (1997) 181–190.
[16] J. Kim, et al., Bone regeneration using hyaluronic acid-based hydrogel with bone morphogenic protein-2 and human mesenchymal stem cells, Biomaterials 28 (10) (2007) 1830–1837.
[17] Z. Shi, et al., Titanium with surface-grafted dextran and immobilized bone morphogenetic protein-2 for inhibition of bacterial adhesion and enhancement of osteoblast functions, Tissue Eng. A 15 (2) (2008) 417–426.
[18] D.-W. Lee, et al., Gentamicin and bone morphogenic protein-2 (BMP-2)-delivering heparinized-titanium implant with enhanced antibacterial activity and osteointegration, Bone 50 (4) (2012) 974–982.
[19] J.H. Lee, et al., In vivo evaluation of CaO-SiO2-P2O5-B2O3 glass-ceramics coating on steinman pins, Artif. Organs 37 (7) (2013) 656–662.
[20] C. Susin, et al., Alveolar ridge augmentation using implants coated with recombinant human bone morphogenetic protein-7 (rhBMP-7/rhOP-1): histological observations, J. Clin. Periodontol. 37 (6) (2010) 574–581.
[21] J. Bolander, et al., The combined mechanism of bone morphogenetic protein-and calcium phosphate-induced skeletal tissue formation by human periosteum derived cells, Eur. Cell. Mater. 30 (2016) 11–25.
[22] H. Sun, H.-L. Yang, Calcium phosphate scaffolds combined with bone morphogenetic proteins or mesenchymal stem cells in bone tissue engineering, Chin. Med. J. 128 (8) (2015) 1121.
[23] K. Majid, et al., Biomimetic calcium phosphate coatings as bone morphogenetic protein delivery systems in spinal fusion, Spine J. 11 (6) (2011) 560–567.
[24] P.-H. Chua, et al., Surface functionalization of titanium with hyaluronic acid/chitosan polyelectrolyte multilayers and RGD for promoting osteoblast functions and inhibiting bacterial adhesion, Biomaterials 29 (10) (2008) 1412–1421.
[25] T.-C. Lee, et al., Bone morphogenetic protein gene therapy using a fibrin scaffold for a rabbit spinal-fusion experiment, Neurosurgery 58 (2) (2006) 373–380.
[26] R. Sasisekharan, S. Ernst, G. Venkataraman, On the regulation of fibroblast growth factor activity by heparin-like glycosaminoglycans, Angiogenesis 1 (1) (1997) 45–54.
[27] T. Ishibe, et al., Bone formation on apatite-coated titanium with incorporated BMP-2/heparin in vivo, Oral Surg. Oral Med. Oral Pathol. Oral Radiol. Endod. 108 (6) (2009) 867–875.
[28] S.E. Kim, et al., The effect of immobilization of heparin and bone morphogenic protein-2 (BMP-2) to titanium surfaces on inflammation and osteoblast function, Biomaterials 32 (2) (2011) 366–373.
[29] B. Trajkovski, et al., Intra-operatively customized implant coating strategies for local and controlled drug delivery to bone, Adv. Drug Deliv. Rev. 64 (12) (2012) 1142–1151.
[30] M. Uezono, et al., Hydroxyapatite/collagen nanocomposite-coated titanium rod for achieving rapid osseointegration onto bone surface, J. Biomed. Mater. Res. B Appl. Biomater. 101 (6) (2013) 1031–1038.
[31] M.R. Leedy, et al., Use of chitosan as a bioactive implant coating for bone-implant applications, in: Chitosan for Biomaterials II, Springer, 2011, pp. 129–165.
[32] M.R. Kumar, et al., Chitosan chemistry and pharmaceutical perspectives, Chem. Rev. 104 (12) (2004) 6017–6084.
[33] R. Muzzarelli, et al., Biological activity of chitosan: ultrastructural study, Biomaterials 9 (3) (1988) 247–252.

[34] R. Muzzarelli, et al., Biochemistry, histology and clinical uses of chitins and chitosans in wound healing, EXS 87 (1999) 251–264.
[35] E. Khor, Chitin: Fulfilling a Biomaterials Promise, Elsevier, 2014.
[36] H. Ueno, et al., Accelerating effects of chitosan for healing at early phase of experimental open wound in dogs, Biomaterials 20 (15) (1999) 1407–1414.
[37] R. Muzzarelli, et al., Stimulatory effect on bone formation exerted by a modified chitosan, Biomaterials 15 (13) (1994) 1075–1081.
[38] K. Masuoka, et al., The interaction of chitosan with fibroblast growth factor-2 and its protection from inactivation, Biomaterials 26 (16) (2005) 3277–3284.
[39] I. Amaral, et al., Attachment, spreading and short-term proliferation of human osteoblastic cells cultured on chitosan films with different degrees of acetylation, J. Biomater. Sci. Polym. Ed. 18 (4) (2007) 469–485.
[40] J. Chang, et al., Biological properties of chitosan films with different degree of deacetylation, J. Mater. Sci. Technol. 24 (5) (2008) 700–708.
[41] J.D. Bumgardner, et al., Chitosan: potential use as a bioactive coating for orthopaedic and craniofacial/dental implants, J. Biomater. Sci. Polym. Ed. 14 (5) (2003) 423–438.
[42] A. Fakhry, et al., Chitosan supports the initial attachment and spreading of osteoblasts preferentially over fibroblasts, Biomaterials 25 (11) (2004) 2075–2079.
[43] A. Abarrategi, et al., Gene expression profile on chitosan/rhBMP-2 films: a novel osteoinductive coating for implantable materials, Acta Biomater. 5 (7) (2009) 2633–2646.
[44] S. Bagheri-Khoulenjani, S. Taghizadeh, H. Mirzadeh, An investigation on the short-term biodegradability of chitosan with various molecular weights and degrees of deacetylation, Carbohydr. Polym. 78 (4) (2009) 773–778.
[45] C. Shi, et al., Therapeutic potential of chitosan and its derivatives in regenerative medicine1, J. Surg. Res. 133 (2) (2006) 185–192.
[46] E.A. Bonfante, et al., Physicochemical characterization and in vivo evaluation of amorphous and partially crystalline calcium phosphate coatings fabricated on Ti-6Al-4V implants by the plasma spray method, Int. J. Biomater. 2012 (2012) 603826.
[47] P.J. Manders, J.G. Wolke, J.A. Jansen, Bone response adjacent to calcium phosphate electrostatic spray deposition coated implants: an experimental study in goats, Clin. Oral Implants Res. 17 (5) (2006) 548–553.
[48] T. Albrektsson, Hydroxyapatite-coated implants: a case against their use, J. Oral Maxillofac. Surg. 56 (11) (1998) 1312–1326.
[49] M. Røkkum, et al., Tissue reactions adjacent to well-fixed hydroxyapatite-coated acetabular cups: histopathology of ten specimens retrieved at reoperation after 0.3 to 5.8 years, Bone Joint J. 85 (3) (2003) 440–447.
[50] R. Jimbo, et al., Genetic responses to nanostructured calcium-phosphate-coated implants, J. Dent. Res. 90 (12) (2011) 1422–1427.
[51] R. Jimbo, et al., Nano hydroxyapatite-coated implants improve bone nanomechanical properties, J. Dent. Res. 91 (12) (2012) 1172–1177.
[52] W. Wijesinghe, et al., Preparation of bone-implants by coating hydroxyapatite nanoparticles on self-formed titanium dioxide thin-layers on titanium metal surfaces, Mater. Sci. Eng. C 63 (2016) 172–184.
[53] L.M. Svanborg, et al., The effect of hydroxyapatite nanocrystals on early bone formation surrounding dental implants, Int. J. Oral Maxillofac. Surg. 40 (3) (2011) 308–315.
[54] G. Munir, et al., The pathway to intelligent implants: osteoblast response to nano silicon-doped hydroxyapatite patterning, J. R. Soc. Interface 8 (58) (2011) 678–688.
[55] S. Barkarmo, et al., Nano-hydroxyapatite-coated PEEK implants: a pilot study in rabbit bone, J. Biomed. Mater. Res. A 101 (2) (2013) 465–471.

CHAPTER 7

# Porous Osseoconductive Layering for Enhancement of Osseointegration

## 7.1 INTRODUCTION

Osseoconductive layer has been traditionally achieved by spraying of the bioactive materials to form an amorphous layer on the surface of the implant. If the size of sprayed powder would be in microscale, then the created roughness might be in microscale. Similarly, by spraying of nano-powder particles, nano-roughness coating layer is structured on the implant surface. Bioactive coating could be processed via different methods. Although with optimization of processing parameters, the coating layer with medium level of porosity and medium adhesion strength (5–30 MPa) is achievable, the consistency of coating layer and created pore size might not be effectively under control in the whole area of coating layer. The advantage of controlling the pore size and porosity level of the coated layer is to enhance the conductivity of the osteogenic cells to be trapped in the created pores at the implant surface and the effective formation of the bone tissue inside the porous structure. In other words, controlled and consistent pore size in similar size with bone extracellular matrix and the percentage of surface porosity close to the bone tissue at the bone-implant interface is beneficial for compact integration of the new lamellar bone through the implant surface. Therefore, porous osseoconductive layering has been developed for a significant improvement of the implant osseointegration. In fact, the porous layer with micro or nano-pores would increase adhesion of the osteogenic cells to the implant surface from which the differentiation and proliferation of the osteogenic cells are sharply increased [1–18]. Likewise, the flow of blood into the pores at the surface of the implant would magnify mineralization of the osteogenic cells while increasing the existence of bone morphogenic protein, growth factors, and other osteoblastic genes and proteins inside the pores for more rapid and effective processing of bone remodeling at the interface of the bone implant. The flow of body fluid at the surface of

*Osseointegration of Orthopaedic Implants*
https://doi.org/10.1016/B978-0-12-813384-2.00007-2

the implant would be significantly increased if the pores are interconnected such as what is structured in the bone tissue matrix, particularly in the trabecular bone. Thus, structuring of the 3D porous with interconnective pores was processed to be attached on the surface of the implant. Furthermore, creation of dynamic porous structure and processing it on the implant surface has been introduced as the promising solution to enhance the bone-implant integration, particularly in the revision surgery of the joint replacements. The structured porous layer is normally developed from nondegradable materials such as titanium and tantalum to establish a permanent osseoconductive layer with plenty of fine pores at the surface of the implant. In this chapter, various techniques of porous coating or structuring for the enhancement of osseoconductivity of the implant surface is reviewed.

## 7.2 LOW POROUS SPUTTER COATING

Physical vapor deposition in forming of coating layer would result to create a much thinner layer with higher adhesion strength. In this method, there would be a carrier gas with positive charge to sputter a nano (or even micro or angstrom) particles from the target specimen and bombard them with a high energy level to the substrate with negative electrical charge (here target specimen is made of the material that supposed to be coated and the substrate is the implant). Through this method, a nano-layer of coating material with the high level of consistency is created on the implant surface. However, the level of porosity might not be close to that of cancellous bone as 60%–90%, while its porosity could be close to that of the cortical bone. From this view, it could be highlighted that sputter-coating process would be greatly advantageous for the promotion of osseointegration in the contact interfaces of orthopedic implants with the cortical bone such as intervertebral or vertebral fusion cages (fusion cage implants are in contact with vertebral endplates) and acetabular cup (prosthesis is in contact with acetabulum with compact structure).

## 7.3 3D POROUS STRUCTURING

The higher level of porosity in the created coating layer by plasma spraying method compared to sputter coating in the contact interface of orthopedic implants with cancellous bone such as femoral component of total knee (prosthesis is in contact with metaphyseal cancellous portion of the distal femur). Due to the limitation of plasma-spraying method to control the consistent porosity and pore size, other method such as electron beam melting (EBM) and selective laser melting (SLM) methods have been introduced as an advanced processing method to form a high porous-coating layer with the controlling advantage of porosity characteristics [1, 3, 7, 16, 17]. These two methods are known as rapid prototyping process, at which the powder of coating material is melted by electron or laser beams. Melting of the powder would consolidate the powder particles at the concentrated point of electron or laser beam. The beam is moved according to the design modeling of the structure which would be imported to the EBM or SLM processing machine. While the beam is moved through the powder according to the designed model or structure, the one layer of powder particles is consolidated. When the first layer is consolidated, it becomes dense. The dense electron or laser-beamed consolidated layer is dropped due to its higher density compared to surrounding powder particles. Then the next layer is beamed and this process is continued layer by layer until processing of the whole structure. Through these two processes, porous structure with similar architecture to that of the trabecular bone could be processed with the porosity between 50% and 90% as that of trabecular or cancellous bone [12]. It was found that SLM method would produce the target designed porous structure more accurate compared to EBM method [10, 13], from which the structure with the higher level of porosity and complexity could be better achievable via SLM processing. The pore size in the range of 260–800 μm could be obtained by SLM process and 350–1400 μm is accessible using EBM process. This is while the higher level of porosity is processed by SLM compared to EBM method. It could be mentioned that, due to more accuracy of SLM process in melting and consolidation of the powder particles, thinner thickness of the structured walls is resulted by this method compared to EBM process. For effective processing of SLM and EBM processes, lower powder size of material is used in SLM process compared to EBM process (i.e., 20–50 μm in SLM vs 45–100 μm [3]). SLM and EBM methods are greatly beneficial for fabrication of total porous implants as the final product. Usage of such rapid methods for coating of a porous layer on the implant surface might not be achievable with good adhesion strength. The processed porous structures by SLM and EBM have been reported with a significant improvement of mesenchymal stem cells attachment compared to plasma-sprayed irregular porous structures [1, 7].

Surface porous structuring has been recently utilized for enhancement of osseointegration of orthopedic implants. Various types of implants with different materials would be processed for the formation of a porous titanium layer on their surface. In total joint replacements, such as total knee, total hip, and total shoulder, the implanted components to the bone are coated by titanium (or tantalum) to promote bone in-growth into the implant surface and ultimately better integration of the implant to the bone. It is attempted to establish a high porous (60%–80%) coating layer with micropores close to that of the trabecular bone. However, in order to achieve an effective biological porous layer, other parameters of the trabecular bone matrix or structure are essentially investigated and produced. The porosity of the trabecular bone is in the form of 3D network matrix with connective pores. However, majority of the coating processing is processed in the form of two-dimensional layers with no control on connection of pores of the created porous layer. Consistency of the porosity is the other morphology character of the trabecular bone which is considered in the development of the porous-coating layer. Interconnectivity and consistency of the porous layer could be alternatively achieved by the attachment of a 3D prestructured interconnected and consistence porous structure on the implant surface.

## 7.4 PROCESSING OF POROUS STRUCTURE IN TRABECULAR SHAPE

As highlighted in the patent US20130218288A1, polyurethane foam scaffold would have the interconnectivity and consistency of a porous structure with a close porosity range and pore size than the trabecular bone. Porous structure of the polyurethane foam could be filled by HA to form the HA slurry followed by heating to burn off the polyurethane. The remained HA slurry is in the similar shape of polyurethane porous structure, which then is coated with titanium or tantalum to obtain the final porous structure. The outer surface of the porous structure could be processed (e.g., abrasion process) to expose the HA core layer of the structure matrix. The advantage of HA exposure in-vivo environment is enhancement of the osseoinductivity at the interface of bone implant. Since the HA material is surrounded with Ti or tantalum layer, the biodegradation of HA is carried out in the long-term postimplantation to promote the mineralization of the new bone cells. One of the main requirements of the porous structure is its compressive strength to be in the range of targeted trabecular bone. Therefore, to achieve the higher compressive strength of Ti or tantalum porous structure, processing of the Ti/tantalum through the porous structure of the polyurethane might be implemented to obtain the Ti/tantalum porous structure with higher compressive strength. Achievement of the porous structure with close morphological, mechanical, and biological characteristics to the target trabecular bone is considered as an advanced solution to enhance the long-term osseointegration of the orthopedic implant, which are in contact with the trabecular bone (e.g., press-fit replacements in implantation of total joint replacements). However, attachment of such structure to the implant is somehow challenging and was investigated through the past two decades.

## 7.5 ATTACHMENT OF POROUS STRUCTURE TO THE IMPLANT SURFACE

The implant and porous structure are attached by incorporation of an intermediate layer comprised of titanium powder and an organic binder [e.g., polyvinyl alcohol (PVA), gelatine, glycerine]. As documented in the patent US20050184134A1, 68% of the intermediate layer is Ti and the balance 32% is mixture of 10% binder agent and 90% water. The intermediate layer is also called "binding mixture" [19] due to compounding of metal powder with an binder agent. The compound of binding mixture is sprayed to the porous structure. Then the coated porous structure is placed and heated on the implant under the appropriate amount of clamping pressure (e.g., 200–1200 psi) while the temperature is adjusted at an appropriate value (e.g., between 100°C and 600°C). Clamping pressure and temperature are inversely proportional which necessitates the careful arrangement of bonding process between the porous structure and implant. It must be highlighted that the shape of the porous structure is essentially needed to be similar with the implant surface that is targeted for the attachment of the porous structure (the implant surface may not be flat and might be curved or has some irregularities such as grooves, threads, extended geometrical features, etc.). Placement of the porous structure in heating step (under clamping pressure) would assemble the porous structure to the implant through binding of the mixture to both implant and porous structure. With the contoured formation of the porous structure on the surface of the targeted implant, the diffusion or sintering step is performed as the final attachment process of porous structuring on the implant. Heating step may take 3 h for the removal of organic binder agent and thermal attachment of the existed Ti powder in the binding mixture to the implant and porous structure. In sintering process, the bonded assembly in the heating step is heated in the temperature of 950°C–1100°C for 4 h. At this temperature, the Ti powder is heated near to the melting temperature to diffuse to the implant surface and porous matrix walls from which the implant is diffused to the porous structure ultimately. Two cycle runs during 4 h at the pressure of effective 400 psi would promote effective and consistent (without gap) bonding of the porous structure on the targeted surface of the implant. Likewise, processing of both heating and sintering steps are carried out in an inert environment conditions using inert gases (e.g., argon or helium) or/and in semi vacuum pressure of 1.33 Pa or less.

The attachment process is based on melting of the intermediate layer between the implant and porous structure to infuse the porous layer on the implant surface. This process might have several uncontrolled disaffects on the characteristics of the porous structure. In fact, the high temperature of about 1000°C [20] in attachment processing could cause the alteration of porosity level and pore size to the undesired range. Likewise, the intermediate layer may penetrate through the porous interconnected network and fill the pores, which affect osteogenic cell differentiation and bone in-growth. This limitation of such attachment process could be improved to establish a very thin layer of binder in between of implant and porous structure to preserve the porous characteristic of the porous structure as much as possible. Processing of porous structure with close mechanical and morphological characteristics to that of the trabecular bone has been introduced as the promising solution for cementless total joint replacements. Likewise, the presence of such porous layer on the surface of the prostheses would significantly enhance its integration with the bone cement in cemented total joint replacements for better press-fit replacement through the bone.

## 7.6 ADVANTAGE OF DYNAMIC POROUS STRUCTURE ON THE IMPLANT SURFACE

As an advanced porous structuring of the implant surface, the concept of dynamic porous structuring was introduced (US20130218288A1). The main target indication of the dynamic porous structure is where a considerable gap or distance is existed between the implant and bone. This scenario might be observed in replacement of some prostheses of total joint implants (e.g., replacement of femoral stem into the femur bone, replacement of tibia tray in tibia bone, and replacement of humeral stem into the humerus bone). In such cases, the interior wall of bone cavity in the diaphyseal portion of the long bone (the cavity of diaphyseal portion consists of bone marrow which is covered by thick layer of compact cortical bone) might be strangely rough (Fig. 7.1) from which the implant could not be press fit through the bone. The advantage of dynamic porous structure on the surface of the implant is

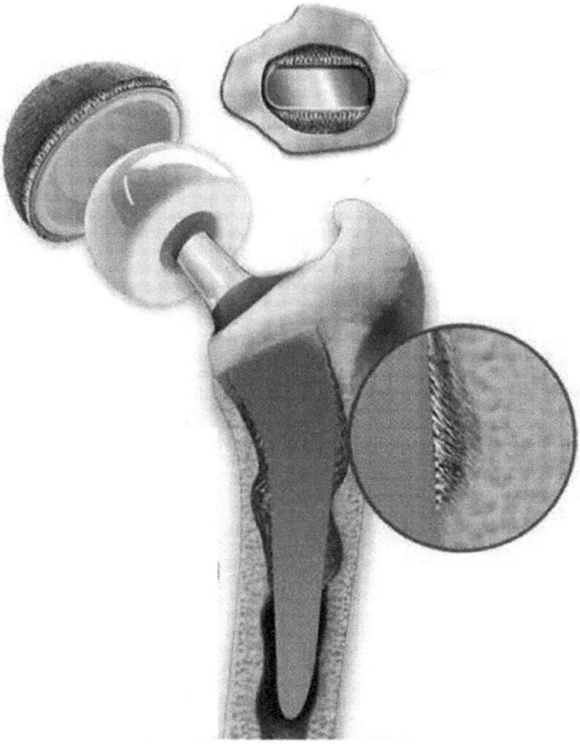

**Fig. 7.1** The rough surface of the diaphyseal proximal femur at the internal cavity; this would affect the effective integration of the implant to the bone, thus by utilization of the dynamic or high elastic porous structure on the surface of the implant or prosthesis, the gap between the bone and prosthesis is entirely filled by effective distribution of porous structure elasticity in accordance with the existed distance between the bone and implant. This would introduce the low amount of preload to conduct the bone in-growth through the attached porous structure on the surface of the prosthesis and facilitate and accelerate osseointegration. *(Courtesy of Fonte et al., Patent no. US20130218288A1.)*

to fill the undesired gaps between the implant and bone. Dynamic structure is achieved by using of nitinol material (nickel titanium) in the fabrication of porous fabric spacer. Nitinol is categorized as shape memory alloy with super elastic characteristics. In Fig. 7.2, the elastic behavior of nitinol is compared with stainless steel and bone. Such super elastic behavior would not have permanent deformation after unloading and would have a large elastic deformation in the high-level range of strain, from which the dynamic porous structured implant is easily fitted and compressed to the bone. Based on the bone profile at the implanted zone, the porous structure is compressed

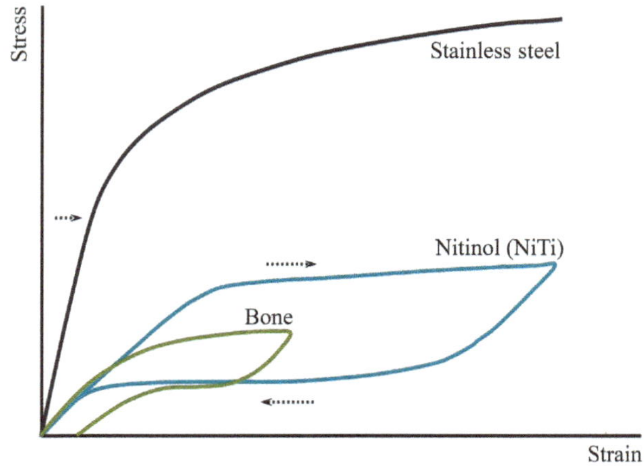


**Fig. 7.2** Schematic illustration of the stress-strain diagram of the stainless steel, nitinol, and bone to distinguish the super elastic behavior of the nitinol with stainless steel and bone, the arrows show the elastic zone; the elastic behavior of the nitinol is observed consistently in both loading and unloading stages which make this material as a super elastic material to be used in the development of orthopedic implants such as a dynamic porous structure layering of the press-fit replacement of the prostheses or implants in long bone cavities.

at different strain values from which the bone-implant gap is entirely filled and generated potential compressive energy would significantly enhance diffusion of bone tissue through the pores of porous structure by which osseointegration is remarkably increased.

## 7.7 DEPTH POROUS STRUCTURING ON THE IMPLANT SURFACE

Creation of porous structure on the surface of the implant has been recently attractive in the development of osseointegration [21–31]. Depth porous structuring has been utilized to enhance the osseointegration of the orthopedic implants that has high press-fit contact with the bone tissue (particularly cortical bone structure such as vertebral endplate or mandibular bone). In high press-fit implantation, the implant is inserted at the targeted zone with impact force. Thus, the implant would bear a permanent value of compressive and shear stresses after implantation. In such situation, coating or creation of a porous layer on the surface of the implant might be associated with wear of the porous layer and releasing of nondegradable particles at the interface of the bone implant. As highlighted in the Section 7.2, coating layer in nanoscale with high adhesion strength could reduce the risk of wear of the coating layer significantly; however, creation of porous layer by laser etching (e.g., on titanium implant) or sulfonating [e.g., on polyetherether-ketone (PEEK) implant] of the surface would be other effective solutions. As following two types of in-depth porous structuring which have been developed to enhance the osseointegration of the orthopedic implants are reviewed in details.

### 7.7.1 Depth Porous Structuring on Titanium Surface

Porous layer could be created in coating thickness of below 5 μm by electrochemical process. In one study, Hen et al. [5] created porous layer of titanium dioxide via anodizing process. They achieved nanoporous $TiO_2$ layer with pore size of about 70 nm on the surface of the PEEK substrate. In their process, the anodizing electrolyte was ethylene glycol ($C_2H_{16}O_2$) containing 2 vol% distilled water and 0.25 wt% ammonium fluoride ($NH_4F$). The process was carried out at the voltage of 60 V for duration of 20 min at field of direct current (DC). The interesting point of this research is that creation of nanoporous coating was managed on the precoated titanium layer via e-beam evaporation. In fact, by deposition rate of 0.05 nm $s^{-1}$, the titanium layer with 2-μm thickness was deposited to the PEEK substrate through the ion beam with the voltage and current of 90 V and 1.5 A, respectively. Anodizing processing of smooth and dense Ti-coated PEEK substrate was resulted to transform the 1.5 mm of the Ti to the $TiO_2$ with pore size of 70 nm (Fig. 7.5). The target application of such porosity is attachment of osteoblastic cells to the implant (e.g., spinal intervertebral fusion cages) in compressed contact with low porous bone tissue (e.g., endplate section of vertebral bone with close bone structure to that of compact cortical bone). Therefore, such porosity would not be adequate for a significant enhancement of osteogenic cells proliferation through the surface of the implant and at the interface of bone-implant. However, for hydrophobic surface (e.g., PEEK polymer), creation of nanoporous layer even with mild interconnectivity would have a significant effect on hydrophilic improvement to allow flow of the blood through the whole interface of the bone-implant. On the other hand, the nanopore size would promote the attachment of osteoblast [32] (the size of osteoblast is about 20–30 μm and could not be trapped through the nanoporous layer, however, the nano-projected arms of the osteoblast cell could be effectively attach through the depth of the holes). In addition, immobilization of the bone morphogenic protein (BMP-2) on the porous $TiO_2$ would be other advantages of nanoporous layer on the surface of the implant (immobilization is immersion of the porous-coated implant to the BMP-2 solution, for instance with concentration of $300\,mg\,mL^{-1}$ [16]. Following immersion, the implants are put into a vacuum desiccator for about 10 min and incubation at 4°C overnight).

### 7.7.2 Depth Porous Structuring of the Polyetheretherketone (PEEK) Surface

Depth porous structuring of the PEEK substrate has been investigated in recent years for application of intervertebral fusion cages. Porous PEEK might be processed via various methods (e.g., extrusion of NaCl crystals, sulfonating). Up to date, only one FDA-approved porous PEEK spinal cage is in the market [33] and thus clinical findings on its effectiveness is scarce. Sulfonating of the PEEK surface for creation of 3D interconnected pores has been developed with promising advantages for enhancement of the PEEK bioactivity in contact surfaces with the bone (Fig. 7.3) [11]. However, the effective mechanical and clinical experiments or observations have not been

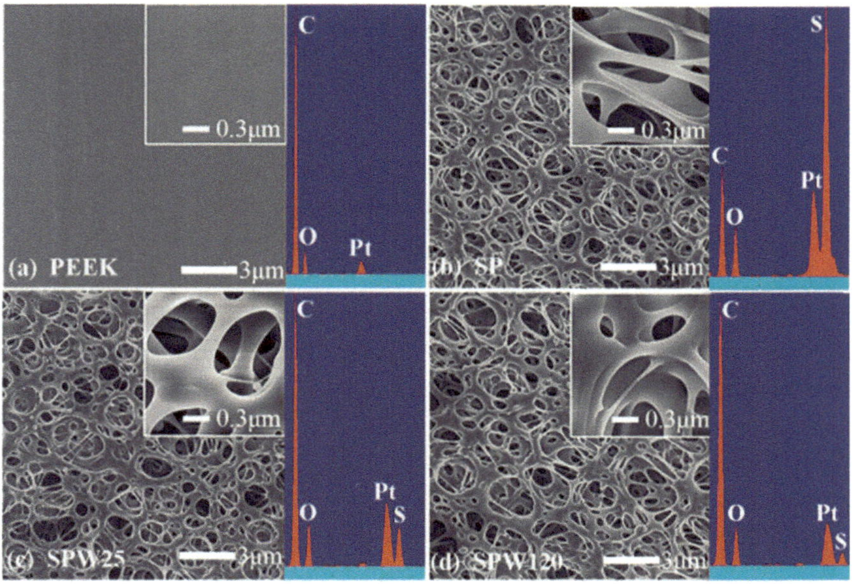


**Fig. 7.3** Illustration of 3D network porous structuring on the surface of the PEEK substrate; this sulfonated surface is processed by utilization of the concentrated sulfuric acid followed by hydrothermal treatment to extract the processing residues from the created 3D interconnected porous layer. *(Courtesy of L. Ouyang, et al., Influence of sulfur content on bone formation and antibacterial ability of sulfonated PEEK, Biomaterials 83 (2016) 115–126 and permission from Elsevier Publisher.)*

published in commercial and clinical indications. Conversely, porous structuring with extrusion of NaCl crystals has been well documented in view of biomechanical and clinical advantages in recent years (since 2016 afterwards).

Since this structuring method has been commercialized for enhancement of the intervertebral fusion cages, further discussion and expression on this technique is provided. In this new structuring, the sodium chloride extrudes the surface of the PEEK to create micro size pores for enhancement of bone in-growth through the PEEK surface (Fig. 7.4). The size of sodium chloride might be different from which various pore sizes could be processed (Fig. 7.5). Porous PEEK would have substantial advantages compared to Ti-coated PEEK or even bare PEEK for intervertebral fusion implantation. There would be some challenging aspects in the development of such new porous structuring conception, which have been highlighted and evaluated in the recent published articles [22, 25, 33–39]. The testing experiments and challenging aspects along with the advantages of the porous PEEK compared to the Ti-coated and bare PEEK cages are discussed in the following. Torstrick et al. [25] investigated the effect of shear strength on the porous surface of the PEEK fusion cage during its impaction to the intervertebral space in

**Fig. 7.4** Illustration of porous structure on the PEEK surface for the enhancement of hydrophilic characteristics of the PEEK in contact with the bone tissue. *(Courtesy of N.T. Evans, et al., High-strength, surface-porous polyether-ether-ketone for load-bearing orthopedic implants, Acta Biomater. 13 (2015) 159–167 and permission from Elsevier Publisher.)*

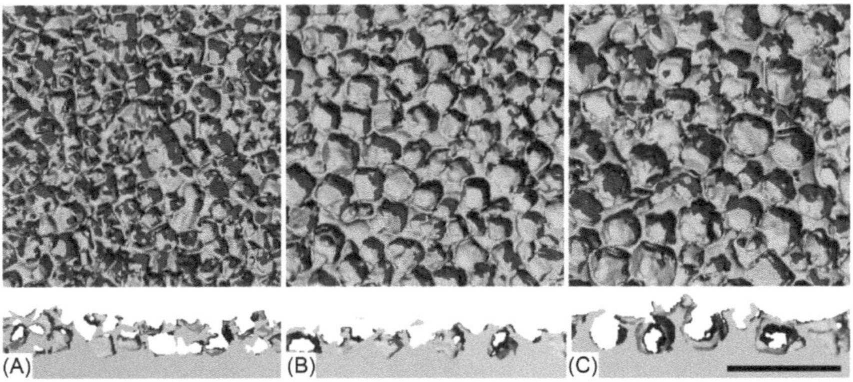


**Fig. 7.5** Extrusion of PEEK surface with sodium chloride crystals in average crystal size of (A) 250 μm, (B) 350 μm, and (C) 450 μm to produce porous structure on the PEEK surface. *(Courtesy of F.B. Torstrick, et al., Do surface porosity and pore size influence mechanical properties and cellular response to PEEK? Clin. Orthop. Relat. Res. 474(11) (2016) 2373–2383 and permission from Springer Publisher.)*

comparison with untreated PEEK (bare PEEK) and titanium-coated PEEK (Ti-coated PEEK) cages. The impaction was performed with 1 lb weight (453.6 g) at a maximum impact speed of $2.6\,ms^{-1}$ to simulate the surgical impaction force for implantation of fusion cage. A constant load of 200 N simulating the physiological load transferring between the vertebral bodies was applied to the solid foam (with density of 40 pcf) blocks (in that research, solid foam blocks are substitutes of spinal vertebrae). It was observed that the porous PEEK cage was placed with one cycle of impaction while the bare PEEK and Ti-coated PEEK cages with almost two and nine impaction cycles, respectively. Likewise, up to 29% of the Ti-coating surface was removed after entire impaction of the cage in between of preload solid foams (Fig. 7.6—it is worth to highlight that the Ti-coating layer was processed by plasma-spraying process). Furthermore, it was reported that the mass change of Ti coated PEEK was significantly higher than that of porous PEEK, which showed higher removal of Ti particles from the coating layer. From the Torstrick's research, superior preservation of characteristics of the treated layer was revealed for porous PEEK compared to Ti-coated implant. One of the differences between the Ti coated and porous PEEK is the porosity level of the treated layer. Porosity of Ti coated has not been well determined in the literature, whereas, the porosity of porous-treated PEEK surface was reported 67% with almost complete interconnectivity. Porosity and interconnectivity of Ti-coating layer might not be accurately measurable due to uncontrollable of porosity level in plasma-spraying and physical

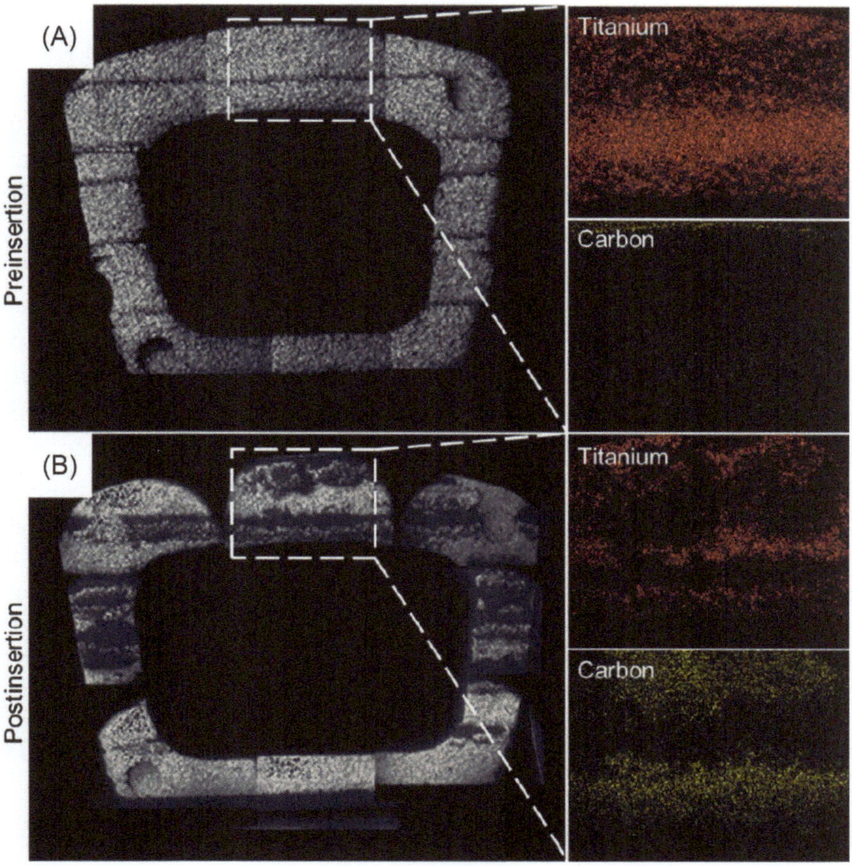


**Fig. 7.6** Removal of Ti particles from the surface of the PEEK cage after entire implantation to the intervertebral space. It is worth to highlight that the Ti-coating layer was processed by plasma-spraying process. Coating processing of Ti with physical vapor deposition (PVD) process may have higher adhesion strength to the PEEK substrate compared to plasma sprayed-coated layer which has not been reported yet. *(Courtesy of F.B. Torstrick, et al., Impaction durability of porous PEEK and titanium-coated PEEK interbody fusion devices, Spine J., (2018) and permission from Elsevier Publisher.)*

vapor deposition (PVD) methods (as current industrial processes for enhancement of the osseointegration in implantation of the PEEK fusion cages). The porosity of these processes is relied on roughness of the implant surface before coating process. One of the challenges of the porous PEEK structuring is the mechanical deterioration, which has been addressed in recent publications [22, 25, 33]. However, this expected deterioration could be eliminated by adequate processing of the porous structuring (with consideration or optimization of the process parameters such as heat, pressure, time of processing, flow distance, etc.) to control the porosity level, pore size, and porous depth. Reduction in tensile strength, interfacial shear strength, ductility, and fatigue strength would be kept negligible for the porous PEEK compared to bare PEEK substitute [22]. This is worth mentioning that the porous layer on the PEEK surface would provide a primary mechanical subsidence of the fusion cage through the vertebral endplate while accelerates the bony fusion between cage and vertebral bodies. Whereas, in normal fusion cages, the primary subsidence is featured by creation of ridges on the PEEK surface and bony in-growth to the depth surface of the cage would not be achieved. Integration of the fusion cage through the bone is totally mechanical interlocking which may cause further penetration of the ridges to the bone over the ridge depth. This excessive penetration would be termed as undesired subsidence of the bone to the fusion cage, which has been contemplated as the clinical parameter in clinical follow-up studies. The undesired subsidence results in reduction of the disc height postoperatively (minor subsidence might be acceptable; however, large amount of subsidence may affect the functionality of spinal column). In contrast, porous layer provides combination of early mechanical and long-term biological integration.

## 7.8 REDUCTION OF STRESS SHIELDING EFFECT IN POROUS SURFACE IMPLANT

Trujillo et al. [26] assessed the balance between stress shielding and bone in-growth in a porous titanium device. Although their results have been reported for a porous component (not porous surface structuring or layering), their concept is discussed here for porous structuring on the implant surface. In general, porous titanium would have much lower mechanical modulus compared to solid titanium (<20 GPa vs 110 GPa) which would result to reduction of stress shielding effect on progression of osseointegration. Porous titanium or tantalum (Ta) structuring on the implant surface would have two advantages of stress shielding reduction and bone in-growth promotion. However, stress-shielding effect is still existed between the Ti or Ta porous structure on the surface of the implant and bulky body of the metallic implant made of titanium alloys, CoCr alloys, or stainless steels.

## 7.9 A DEVELOPMENT IDEA

The porous structure with the decreased level of porosity from the surface through the depth of the structure might be an alternative solution to reduce the effect of stress shielding gradually at the interface of the bone implant. This would be a novel concept to enhance the long-term osseointegration with the improvement of load-transferring conditions through the porous layer. This enhanced porous layer might be processed in higher thickness than the porous layer with one level of porosity. Different level of porosity would cause various pore sizes, which generates multiple levels of strains. The required range of strain for the stimulation of bone remodeling process and flow of blood in to the porous layer would be expected to be more effective in multiporous layer compared to mono-porous layer. The reason is the existence of a range of strain values, which would cause acceleration of the bone remodeling process and generate a regime of strain distribution through the depth of multiporosity layer. The strain range would be lower near to the bulky body of the implant at which the bone remodeling might be more accelerated through the depth of the multiporous layer. By formation of the lamellar bone at the deep depth of the porous layer, the strain range in the upper line becomes lower and stimulates the required mechanical signal for the induction of osteogenic cells to generate collagen fibrils and mineralization of the created osteocyte cells.

## 7.10 DISCUSSION ON SOME CHALLENGING ASPECTS OF POROUS LAYERING

Porous structuring on the surface of the orthopedic implant has been found as a key concept for a significant enhancement of the bone-implant integration at the contact interface. However, various mechanical and biological concepts might be crucial to develop an effective porous surface layer. Biologically, hypoxia effect has been addressed as the side effect of porous layering at the bone-implant interface [33, 40–42]. The concern is nondiffusion of nutrient through the deeper section of the porous layer which would be recognized by increasing the expression of vascular endothelial growth factors (VEGFs) in osteoblast cell culture (MC3T3) [22]. However, it was revealed that blood vessels are able to perfuse through the interconnective pores of the porous structure until the deepest pores facilitate vascularization to prevent hypoxia effect [33]. The other concern of porous structure on the surface is to increase the risk of deterioration in the fatigue strength due to increasing of the risk of crack initiation points at the porous structure and its propagation through the solid body of the implant. In fact, such reduction of fatigue strength might be accelerated in the axial shear (generated by local axial compression or bending stresses) stress. For instance, in the case of PEEK porous implant, if the level of effective stress on the porous layer would be less than 38 MPa (endurance limit or fatigue strength of PEEK), risk of crack initiation is eliminated. For instance, the lumbar intervertebral fusion cage would normally bear the load of 1000–3000 N depending on the level of movement and load-bearing condition of the spinal column in various activities [43, 44]. With consideration of 3000 N as the maximum lumbar intervertebral load, at least a contact surface of 80 $mm^2$ is required to maintain the stress level less than 38 MPa. This geometrical point must be considered during the development of lumbar intervertebral fusion cage to reduce the risk of crack creation on the processed porous layer of the cage. Such technical analyses are substantially required to enhance the effectiveness of the porous structuring on the surface of the bioinert orthopedic implants. Low level of mechanical vibration has been found effective in the enhancement of the osteogenesis process at the interface of the bone and the implant with porous surface, from which osseointegration of the implant is promoted [45, 46].

## 7.11 SUMMARY

Porous structuring or layering on the surface of the orthopedic implants has been found to be the superior method of the enhancement of the bioinert implant bioactivity in contact surfaces with the bone tissue, particularly the trabecular bone. The surface of the implant could be coated by osseoconductive materials such as titanium or tantalum in low porosity level (less than 30%) to be compatible with compact cortical bone or processed with the high level of porosity to be compatible with the trabecular bone. Both porous layering methods are under interest for promotion of bone in-growth through the surface of the implant for long-term integration of the bone tissue to the implant. The porous layer is processed in nano, micro, or even macrolevel with various advantages and benefits. The nanoporous layer (pore size between 1 and 500 nm) would enhance the adhesion of the osteogenic cells, morphogenic proteins, growth factors, osteoblastic genes, and other biological factors while the microporous layer (pore size between 500 nm and 100 μm) would enhance lining and proliferation of the osteoblast cells at the interface of the bone and implant. Furthermore, the macroporous layer (pore size between 100 μm and 1000 μm) would enhance differentiation of the osteoblast cells to the osteocyte cells followed by mineralization and maturation of the new bone cells to be formed as the lamellar bone (the provided classification of porous sizes would be based on the practical experiences of the author which might be different with some literature). In fact, for effective osseointegration of the bone to the implant, all these advantages and benefits are required and carried out during the bone remodeling process at the interface of the bone and the implant nevertheless of the level of the porosity of the created porous layer. However, with consideration of adequate level of porosity in accordance with the intended indication, the porous layer is processed on the surface of the implant. For instance, in order to enhance the bioactivity of the PEEK surface in intervertebral fusion implantation, it is crucial to form a nano-pores layer with a low-level porosity to promote the adhesion strength of the porous layer to the PEEK cage during implantation and to enhance the adhesion of the biological factors to the nano-pores of the porous layer for acceleration of the bone remodeling process. Otherwise, the induction of the osteogenic cells might not be achieved with the formation of the micro- or macro-porous layer for this indication. In this chapter, the formation of various types of porous layering on the surface of the orthopedic implant in accordance with their intended indications has been reviewed and their advantages and challenging aspects were discussed. 3D porous structuring on the surface of the implant in contact with the trabecular bone with high level of porosity (more than 70%) could be achieved by processing of the 3D structure separately and then bind or attach it to the implants surface. This 3D structure could be fabricated from the super elastic alloys (e.g., nitinol) to improve the osseointegration of the press-fit prosthesis to the affected cavities of the long bone in total joint replacement. The challenging aspects of the porous structure binding to the implant and advantages of the 3D porosity have been deeply discussed in Sections 7.4–7.6. Likewise, the concept of depth porous structuring has been reviewed and discussed for two types of utilized method for enhancement of PEEK bioactivity characteristics in Section 7.7. At the final section of this chapter, the effect of porous layering on stress shielding at the bone-implant interface is generally highlighted. Accordingly, a novel concept was hypothesized to reduce the effect of stress shielding for increasing of the effectiveness of the porous layer in the long-term implantation. A brief discussion was provided on other aspects of the porous layering such as hypoxia effect and reduction of fatigue strength. Overall, the porous structuring or layering on the surface of the orthopedic implant has been revealed as the advanced solution for the long-term osseointegration of the implant, which is now under technological development to be commercialized and utilized in the treatment of bone injuries.

## 7.12 REMIND AND LEARN

**Q7.1** What are the general advantages of the porous osseoconductive layering compared to the bioactive osseoconductive layering?

**Q7.2** Describe sputter-coating process for the formation of the osseoconductive layer on the implant surface. What are the characteristics of the sputter-coated layer?

**Q7.3** What clinical indications might be advantageous to be treated by forming of the sputtered layer on the implant surface?

**Q7.4** What is the main advantage of EBM, and SLM methods compared to the plasma-spraying method?

**Q7.5** Describe the process of EBM, and SLM methods for fabricating of the 3D porous structure. Compare the effectiveness of these two methods to process the porous structure. Express the usability of the SLM and EBM methods in the development of orthopedic implants.

**Q7.6** What materials are adequate for fabricating of the 3D porous structure? Describe which morphology characteristics of the trabecular bone would be utilized for the development of the surface porous structuring.

**Q7.7** Express the processing of the porous structure in the shape of the trabecular bone according to the US20130218288A1.

**Q7.8** How the porous structure might be attached to the implant surface? Describe the processing method according to the patent US20050184134A1. What processing parameters are crucial for the attachment of porous structure to the implant surface?

**Q7.9** What are the heating and sintering steps in the attachment of porous structure to the implant surface? What is the range of temperature in these two steps? What environment conditions might be implemented for processing of these two steps?

**Q7.10** What are the disaffects of the binding process in the attachment of the porous structure to the implant surface?

**Q7.11** Express the concept of dynamic porous structuring on the surface of the implant. What indications might be treated by this conception?

**Q7.12** What type of material is adequate to be used for the fabrication of dynamic porous structure? How is the behavior of this material during implantation to the bone? Discuss how this behavior would enhance the osseointegration remarkably.

**Q7.13** What is the meaning of depth porous structuring? Compare the characteristics of this type of surface structuring with porous coating.

**Q7.14** What is the main disadvantage of coating or creation of a porous layer on the surface of the implant in press-fit implantation with great impaction force?

**Q7.15** Express the process of nanoporous $TiO_2$ structuring on the surface of the PEEK substrate. What is the limitation of the created structure in view of the bioactivity? What is the advantage of such processing in enhancement of hydrophilic characteristics of the PEEK surface? How about the advantage of immobilization of the bone morphogenic protein to the PEEK surface?

**Q7.16** Express the depth porous structuring of the PEEK surface. What mechanical characteristics of this porous structuring would be crucial to be evaluated?

**Q7.17** What is the impaction testing method for the evaluation of porous PEEK fusion cages? What is the comparison result of the impaction test between porous PEEK, Ti-coated PEEK, and bare PEEK cages?

**Q7.18** Discuss the mechanical deterioration of the porous PEEK compared to Ti-coated PEEK and bare PEEK fusion cages.

**Q7.19** Discuss the advantages of porous PEEK cage compared to cages with the ridges in view of the primary and secondary subsidence of the cage. How about bony fusion at the interface of bone with porous PEEK, Ti-coated PEEK, and bare PEEK?

**Q7.20** Discuss the effect of porous structuring on the extent of stress shielding at the bone-implant interface.

**Q7.21** Describe the novel concept of multiporous structuring on the surface of the implant and discuss its advantages on the enhancement of the osseointegration.

## 7.13 THINK AND DISCUSS

**Q7.22** Discuss the advantages and disadvantages of the depth porous structuring compared to the porous layering by deposition or coating processing methods.

**Q7.23** Discuss how the hypothesized idea in Section 7.9 could be developed for commercial use.

**Q7.24** What other porous structuring methods could be hypothesized for the enhancement of the long-term osseointegration between the implant and the bone tissue.

## REFERENCES


[1] J.E. Biemond, et al., Assessment of bone ingrowth potential of biomimetic hydroxyapatite and brushite coated porous e-beam structures, J. Mater. Sci. Mater. Med. 22 (4) (2011) 917–925.

[2] J.A. Hanzlik, J.S. Day, Bone ingrowth in well-fixed retrieved porous tantalum implants, J. Arthroplast. 28 (6) (2013) 922–927.

[3] J. Biemond, et al., Bone ingrowth potential of electron beam and selective laser melting produced trabecular-like implant surfaces with and without a biomimetic coating, J. Mater. Sci. Mater. Med. 24 (3) (2013) 745–753.

[4] J. Matena, et al., Comparison of selective laser melted titanium and magnesium implants coated with PCL, Int. J. Mol. Sci. 16 (6) (2015) 13287–13301.

[5] C.M. Han, et al., Creation of nanoporous TiO2 surface onto polyetheretherketone for effective immobilization and delivery of bone morphogenetic protein, J. Biomed. Mater. Res. A 102 (3) (2014) 793–800.

[6] M. Ogawa, et al., Early fixation of cobalt-chromium based alloy surgical implants to bone using a tissue-engineering approach, Int. J. Mol. Sci. 13 (5) (2012) 5528–5541.

[7] J.E. Biemond, et al., The effect of e-beam engineered surface structures on attachment, proliferation and differentiation of human mesenchymal stem cells, Biomed. Mater. Eng. 21 (5-6) (2011) 271–279.

[8] S. Kujala, et al., Effect of porosity on the osteointegration and bone ingrowth of a weight-bearing nickel–titanium bone graft substitute, Biomaterials 24 (25) (2003) 4691–4697.

[9] N. Bertollo, et al., Effect of surgical fit on integration of cancellous bone and implant cortical bone shear strength for a porous titanium, J. Arthroplast. 26 (7) (2011) 1000–1007.

[10] R. Van Noort, The future of dental devices is digital, Dent. Mater. 28 (1) (2012) 3–12.

[11] L. Ouyang, et al., Influence of sulfur content on bone formation and antibacterial ability of sulfonated PEEK, Biomaterials 83 (2016) 115–126.

[12] L. Murr, et al., Microstructure and mechanical behavior of Ti–6Al–4V produced by rapid-layer manufacturing, for biomedical applications, J. Mech. Behav. Biomed. Mater. 2 (1) (2009) 20–32.

[13] A. Fukuda, et al., Osteoinduction of porous Ti implants with a channel structure fabricated by selective laser melting, Acta Biomater. 7 (5) (2011) 2327–2336.

[14] F. Matassi, et al., Porous metal for orthopedics implants, Clin. Cases Miner. Bone Metab. 10 (2) (2013) 111.

[15] H. Shahali, A. Jaggessar, P.K. Yarlagadda, Recent advances in manufacturing and surface modification of Titanium orthopaedic applications, Procedia Eng. 174 (2017) 1067–1076.

[16] A. Agapovichev, et al., Selective laser melting of titanium alloy: Investigation of mechanical properties and microstructure, in: IOP Conference Series: Materials Science and Engineering, IOP Publishing, 2016.

[17] J. Matena, et al., SLM produced porous titanium implant improvements for enhanced vascularization and osteoblast seeding, Int. J. Mol. Sci. 16 (4) (2015) 7478–7492.

[18] D. Cristea, I. Ghiuta, D. Munteanu, Tantalum based materials for implants and prostheses applications, Bull. Transilv. Univ. Brasov. Eng. Sci. Ser. I **8** (2) (2015) 151.

[19] S. Charlebois, et al., Method for Attaching a Porous Metal Layer to a Metal Substrate, Google Patents, 2017.

[20] M. Fonte, M. Palmer, Dynamic Porous Coating for Orthopedic Implant, Google Patents, 2015.

[21] Frauchiger, V.M., et al., An open-porous titanium coating for advanced osseointegration, in: Proc. 55th Annual meeting orthopaedic research society.

[22] F.B. Torstrick, et al., Do surface porosity and pore size influence mechanical properties and cellular response to PEEK? Clin. Orthop. Relat. Res. **474** (11) (2016) 2373–2383.
[23] D. Wang, et al., Coatings for osseointegration of metallic biomaterials, in: Surface Coating and Modification of Metallic Biomaterials, Elsevier, 2015, pp. 345–358.
[24] N. Taniguchi, et al., Effect of pore size on bone ingrowth into porous titanium implants fabricated by additive manufacturing: an in vivo experiment, Mater. Sci. Eng. C 59 (2016) 690–701.
[25] F.B. Torstrick, et al., Impaction durability of porous PEEK and titanium-coated PEEK interbody fusion devices, Spine J. **18** (5) (2018) 857–865.
[26] C. Domínguez-Trujillo, et al., Improvement of the balance between a reduced stress shielding and bone ingrowth by bioactive coatings onto porous titanium substrates, Surf. Coat. Technol. (**338**, 2018, 32–37).
[27] Y. Ren, et al., Microwave assisted coating of bioactive amorphous magnesium phosphate (AMP) on polyetheretherketone (PEEK), Mater. Sci. Eng. C 85 (2018) 107–113.
[28] A. Rubshtein, et al., Porous material based on spongy titanium granules: structure, mechanical properties, and osseointegration, Mater. Sci. Eng. C 35 (2014) 363–369.
[29] C. Domínguez-Trujillo, et al., Sol-gel deposition of hydroxyapatite coatings on porous titanium for biomedical applications, Surf. Coat. Technol. 333 (2018) 158–162.
[30] H. Mahjoubi, et al., Surface phosphonation enhances hydroxyapatite coating adhesion on polyetheretherketone and its osseointegration potential, Acta Biomater. 47 (2017) 149–158.
[31] Y. Zhao, et al., Cytocompatibility, osseointegration, and bioactivity of three-dimensional porous and nanostructured network on polyetheretherketone, Biomaterials 34 (37) (2013) 9264–9277.
[32] W.Q. Yu, et al., The effect of anatase TiO2 nanotube layers on MC3T3-E1 preosteoblast adhesion, proliferation, and differentiation, J. Biomed. Mater. Res. A **94** (4) (2010) 1012–1022.
[33] N.T. Evans, et al., High-strength, surface-porous polyether-ether-ketone for load-bearing orthopedic implants, Acta Biomater. 13 (2015) 159–167.
[34] H. Ahn, et al., Biological evaluation and finite-element modeling of porous poly (para-phenylene) for orthopaedic implants, Acta Biomater. 72 (2018) 352–361.
[35] N.T. Evans, et al., Local deformation behavior of surface porous polyether-ether-ketone, J. Mech. Behav. Biomed. Mater. **65** (2017) 522–532.
[36] N.T. Evans, et al., Impact of surface porosity and topography on the mechanical behavior of high strength biomedical polymers, J. Mech. Behav. Biomed. Mater. 59 (2016) 459–473.
[37] R.D. Carpenter, et al., Effect of porous orthopaedic implant material and structure on load sharing with simulated bone ingrowth: a finite element analysis comparing titanium and PEEK, J. Mech. Behav. Biomed. Mater. 80 (2018) 68–76.
[38] F.B. Torstrick, et al., Getting PEEK to stick to bone: the development of porous PEEK for interbody fusion devices, Tech. Orthop. 32 (3) (2017) 158–166.
[39] K. Burkus, Early outcomes of anterior cervical discectomy and fusion using a porous PEEK interbody fusion device. J. Spine Neurosurg. **7** (2) (2018) https://doi.org/10.4172/2325-9701.1000295.
[40] J. Dai, A. Rabie, VEGF: an essential mediator of both angiogenesis and endochondral ossification, J. Dent. Res. 86 (10) (2007) 937–950.
[41] H. Mayer, et al., Vascular endothelial growth factor (VEGF-A) expression in human mesenchymal stem cells: autocrine and paracrine role on osteoblastic and endothelial differentiation, J. Cell. Biochem. 95 (4) (2005) 827–839.
[42] D.S. Steinbrech, et al., VEGF expression in an osteoblast-like cell line is regulated by a hypoxia response mechanism, Am. J. Phys. Cell Phys. 278 (4) (2000) C853–C860.

[43] A.B. Schultz, G.B. Andersson, Analysis of loads on the lumbar spine, Spine 6 (1) (1981) 76–82.
[44] H.J. Wilke, et al., New in vivo measurements of pressures in the intervertebral disc in daily life, Spine 24 (8) (1999) 755–762.
[45] D. Jing, et al., Effect of low-level mechanical vibration on osteogenesis and osseointegration of porous titanium implants in the repair of long bone defects, Sci. Rep. 5 (2015) 17134.
[46] D. Jing, et al., Pulsed electromagnetic fields promote osteogenesis and osseointegration of porous titanium implants in bone defect repair through a Wnt/β-catenin signaling-associated mechanism, Sci. Rep. 6 (2016) 32045.

CHAPTER 8

# Conception of Osseointegration in Development of Orthopedic Implant Industry

## 8.1 INTRODUCTION

Due to the enhancement of technology in various manufacturing and finishing processes, the development of orthopedic implants would be considerably pursued by the incorporation of new technological discoveries. These new achievements would cause introduction of new biomaterials, new featuring technologies, new manufacturing methods, new finishing processing, and new design structuring. In orthopedic implants, all these development aspects would generate various types of strategies to enhance the biomechanical, biological, and clinical benefits of the surgical treatment according to the growth and complexity of the clinical needs. There are multiple development concepts associated with orthopedic implants that are considered as the development strategies, which are illustrated in Table 8.1.

From the data in Table 8.1, it can be observed that the strategies related to the bone formation (Nos. 4 and 9) are enhancing all biomechanical, clinical, and biological requirements of the surgical treatment. Indeed, bone-implant integration would enhance the biomechanical, biological, and clinical benefits of the implant after implantation. The concept of bone-implant integration would be mechanically and biologically discussed. Traditionally, mechanical integration of the implant to the bone was the main target of surgical treatment of severe bone injuries. Gradually, biological interaction between the implant and the bone was detected to be substantially crucial for the long-term safety and performance of the implant. In this chapter, the concept of osseointegration from mechanical and biological aspects is discussed and then the engineering and evaluation concerns are expressed.

*Osseointegration of Orthopaedic Implants*
https://doi.org/10.1016/B978-0-12-813384-2.00008-4

**Table 8.1** Development strategies or concepts in orthopedic implant industry

| No. | Strategy | Development item | Type of strategy |
|---|---|---|---|
| 1 | Enhanced design features | Implant geometry, implant structure | Clinical and biomechanical needs |
| 2 | Enhanced and easier replacement | Surgical instruments, implant variety | Clinical needs |
| 3 | Biomechanical stability, mechanical integration | Implant-bone stiffness, implant- bone assembly fixation | Biomechanical needs |
| 4 | Biological stability, biological integration | Implant-bone integration, bone in-growth and on-growth through the implant surface | Biological, clinical, and biomechanical needs |
| 5 | Mechanical flexibility | Implant-bone optimal stiffness, implant-bone assembly with micro-dynamic fixation | Biomechanical and biological needs |
| 6 | Soft tissue preservation | Implant design, implant structure, implant technology | Biomechanical and clinical needs |
| 7 | Preservation of motion ranges | Implant design, implant structure, implant technology | Biomechanical and clinical needs |
| 8 | Reduced wear | Surface molecular bonding, surface hardness, surface coating, contact stress, sliding distance | Biomechanical |
| 9 | Bony fusion | Bone to bone contact stability, intermediate osseoconductive structure | Biological, clinical, and biomechanical needs |
| 10 | Drug delivery | Diffusion of growth factors, proteins, etc. to the surface | Biological |
| 11 | Antibacterial featuring | Bacterial surface adherence, drug immobilization on the surface | Clinical and biological needs |

## 8.2 MECHANICAL BONE-IMPLANT INTEGRATION

The integration of the bone to the implant or implant to the bone has been the attractive topic for researchers, scientists, manufacturers, and even specialist clinicians. Although, there are plenty of challenges and concerns in osseointegration projects, its superior beneficial advantages were revealed by all the implant developers. Traditionally, orthopedic implants were developed from bioinert metals such as stainless steel and gradually promoted to the titanium alloys implants. The first titanium implant was manufactured as dental implant to be implanted into the mandibular bone as the base structure for artificial crown. With the growth of technology in material engineering of metallic alloys and polymers, other kinds of materials such as cobalt chromium (CoCr) alloys, ultrahigh-molecular-weight polyethylene (UHMWPE) polymer, and polyetheretherketone (PEEK) polymer were utilized in the development of the orthopedic implants. The main concern of using such bioinert materials is poor establishment of the implant-bone integration which increases the risk of instability of the implant at implantation site. Due to high mechanical properties of the bone tissue (particularly at the cortical layer of the bones), mechanical integration of the bioinert implants to the bone could be achieved in orthopedic implants that are implanted by purchasing or tapping of the well-designed screws. However, in some cases, mechanical integration might not be established in early or long-term implantation (placement of the total joint prostheses such as knee femoral, shoulder glenoid cup, intervertebral/vertebral fusion cages, or hip acetabular cup to the bone are the examples that might be managed with poor mechanical integration to the bone). Likewise, mechanical integration of the implant through the bone with low mineral density might not be achieved effectively. Loosening of the implant might have been caused due to poor mechanical bone-implant integration (the mechanical aspects of implant loosening were reviewed and discussed in Chapter 1). Therefore, in parallel with the enhancement of load transferring at the bone-implant interface, biological integration was addressed as significant factor in the promotion of implant osseointegration. It is worth mentioning that both mechanical and biological development concepts are highly effective when contemplated together.

## 8.3 COMBINED MECHANICAL, BIOLOGICAL, AND CLINICAL FACTORS IN BONE-IMPLANT INTEGRATION

The effective physical and mechanical parameters at the bone-implant interface are established by the implant design, methods of manufacturing, methods of postproduction processes, methods of surface finishing, structure of the implant (screw, flat, or curve-shaped structure), mechanism of implant application when developed as assembly, material properties (physical and mechanical properties such as flexural strength, shear strength, fatigue strength, wear resistant, density, surface hardness, surface morphology, etc.), and the level of rigidity/flexibility of the bone-implant assembly. On the other hand, the influence of biological parameters which are involved in bone remodeling process at the bone-implant interface would be crucial in effective and long-term osseointegration of the orthopedic implant at the implanted site (the involved biological factors could be known as osteogenic cells, osteoblast cells, growth factors, osteoblastic genes, proteins, etc.). In fact, effective biological performance of the new bone formation and old bone resorption after implantation of the implant would be a biological effect that enhances the osseointegration of the implant. Other biological effects such as bacterial adhesion and fibrous layering between the implant and the bone might be occurred at the bone-implant interface. Furthermore, there are several clinical requirements that are considered in the enhancement of bone-implant integration. For instance, screw purchasing in cancellous and cortical bones could be considered as a clinical parameter which causes different designing and structuring of the screw threads.

Consideration of all mechanical, biological, and clinical parameters is absolutely essential for effective bone-implant integration and is beneficial to be considered in the development stage of the orthopedic implants. In other words, when a biological factor of the implant is enhanced, other mechanical and clinical factors of the implant would be crucial to be studied or evaluated to assure the effectiveness of the implant osseointegration. For instance, as highlighted in Chapter 7, porous PEEK intervertebral fusion cage was developed to enhance its biological integration to the vertebral endplates, however, rigorous mechanical evaluations were carried out to examine the reduction in mechanical properties of the porous surface compared to the cage surface with ridges. Other example is using of the titanium alloys for manufacturing of the orthopedic implants, which is a mechanical consideration for better load-bearing of physiological loading conditions, while increasing the risk of stress-shielding effect and undesired bone resorption at the bone-implant interface. More examples are illustrated in Table 8.2 to provide effective expression of the conflict between biological and mechanical factors in designing and developing of orthopedic implants.

This confliction would be inevitable which would create a new field of multidiscipline science involving biology, medicine, material, mechanics, chemistry, and physics sciences to optimize the biological and mechanical benefits of the implant integration to the bone. However, based on the application of the implant, one of the mechanical or biological benefit might

**Table 8.2** Conflicts between biological and mechanical factors in designing and development of orthopedic implants

| No. | Type of implant | Biological-mechanical conflict |
|---|---|---|
| 1 | Screw thread depth | Depth of the screw thread at the screw shaft would increase tapping and purchasing of the screw to the bone from which the biological integration of the bone to the screw is significantly increased. This biological benefit may reduce the core diameter of the screw shaft and thus deduction of the flexural strength of the screw against compression or bending stresses. |
| 2 | Porous structure attachment to implant surface | Attachment of porous structure would significantly enhance the bone in-growth or penetration of the bone cement through the implant surface in total joint replacements. However, such biological benefit may not be suitable in some types of implants such as spinal intervertebral fusion cages at which the porous structure might not be mechanically resistant to impaction-shear force which is applied during implantation stage. |
| 3 | Bioactive coating of trauma screw | Bioactive coating of screws which are used for fracture fixation of trauma is an advantage for better integration of the screw to the bone and ultimately fixation stability. However, the screw might not be removed after completion of the treatment period or in revision surgeries easily. Even it may cause cracking of the bone at the implanted site which is clinically catastrophic. |
| 4 | Implant surface blasting | Blasting of the implant surface would increase the surface roughness and adhesion of osteogenic cells and other biological factors to the implant surface from which osseointegration is enhanced. However, blasting of the surface may cause creation of microcracks on the surface of the implant, thus reducing the fatigue strength of implant. |

*Continued*

**Table 8.2** Conflicts between biological and mechanical factors in designing and development of orthopedic implants—cont'd

| No. | Type of implant | Biological-mechanical conflict |
|---|---|---|
| 5 | Ti coating of Spinal PEEK cage | Ti coating of spinal intervertebral PEEK cages would enhance hydrophilic characteristics of the PEEK surface form which adhesion of osteogenic cells and other biological factors to the implant surface is increased. However, implantation of the PEEK cage by impaction force may cause detachment of the Ti particles at the bone-implant interface and ultimately bone resorption at the interface. |
| 6 | Cementless total joint prostheses | Cementless total joint prosthesis are enhanced by coating of bioactive ceramics on the connective surfaces with the bone to provide biological integration of the prosthesis with the bone compared to the bone-implant integration with bone cement. However, due to the high level of loading conditions at the interface of the bone and prosthesis, bone remodeling process to form new mature bone at the interface might not be effectively established. |
| 8 | Spinal PEEK cage with ridges | Spinal cages with ridges would increase the mechanical integration of the vertebral body to the cage. However, this integration may cause over subsidence of the vertebrae to the cage, thus reducing the disc height and changing the relative anatomical positioning of the involved vertebrae. |
| 9 | Highly rigid spinal stabilization | Highly rigid spinal stabilization may affect on fusion process of two adjacent vertebrae to the degenerated disc. Existence of micromotion at the stabilized site would enhance the stimulation of the required mechanical factors for promotion of the bone remodeling process at the interface of vertebrae and bone graft from which the fusion or bone formation is enhanced. |

**Table 8.2** Conflicts between biological and mechanical factors in designing and development of orthopedic implants—cont'd

| No. | Type of implant | Biological-mechanical conflict |
|---|---|---|
| 10 | Rigid trauma fixation | Rigid trauma fracture fixation may affect on the healing process of bone fragments. Existence of micromotion at the stabilized or fixed site would enhance the stimulation of the required mechanical factors for promotion of bone formation at the gaps of bone fragments from which the bone healing is enhanced. |

be prominent, for instance, utilization of high-strength trauma plating system [made of titanium (Ti)] for fixation of pelvic fractures. In such situations, one aspect of osseointegration is targeted which necessitates no consideration of other aspect as the priority factor.

## 8.4 MECHANICAL INTERLOCKING BETWEEN THE BONE AND THE IMPLANT

One of the concepts that influences osseointegration conception is how to look into the bone. Among the body tissues of the human body, bone tissue is hard which has high mechanical load-bearing characteristics. Like other body tissues, bone tissue is a smart bioactive structure that is alive which means that the dead cells are resorbed and new cells form new bone tissue to substitute bone loss. In view of orthopedic implant development, mechanical properties of the bone have been prominently focused. In fact, the implant could be implanted into the bone to form a stable bone-implant construct. Concerning implant osseointegration, biological characteristics of the bone tissue are crucial but still not prominent. There would be essential need for bone generation and resorption at the bone-implant interface, however, mechanical interlocking of the implant to the bone is more prominent for early and long-term stability of the implant in contact with the bone. This interlocking is provided by internal (e.g., screw threads) and external (e.g., extended tab or pins in knee femoral component) features in macroscale. The risk of such interlocking is loosening of the implant postoperatively after couple of months or years (implant loosening was reported in majority of the clinical publications which were reviewed in Chapter 1). Since the bone remodeling process is accomplished in microscale, the mechanical interlocking could be obtained by micro- or nanostructuring of the implant surface (as reviewed in Chapters 6 and 7). This could establish an optimal mechanical-biological condition for more effective and long-term osseointegration.

## 8.5 CLINICAL ASPECTS OF BONE-IMPLANT INTEGRATION

One of the key factors to achieve the effective bone-implant integration is adequate intraoperative surgical techniques and clinical postoperative treatment plans. It is crucial to minimize the opening of the operated site by the incorporation of advanced surgical instruments and by the use of surgical instrumentation methods (e.g., minimal invasive surgery–MISS) which would reduce the risk of infection and soft tissue complications after operation (e.g., nonhealing of incision site, disorder of muscles contraction and extension, disorder of ligament and tendon tensioning, irritation of skin, ligament, tendon, or muscle after implantation). Likewise, postoperative physiotherapy plan is highly effective in preventing overloading and less loading of the implantation site. In fact, by further controlling of the clinical aspects, the effective bone-implant integration is more manageable. Therefore, nowadays, the implant developers and manufacturers are developing implants along with instrumentation and posttreatment techniques and solutions. Today, the importance of surgical instruments is recognized to some extent that the well-established manufacturers have their own instrument sets in accordance with the specific features and benefits of the developed implant. This is because the techniques of bone preparation, soft tissue preservation, minimal soft tissue irritation and suppression, maximal intact bone preservation, accurate implant size determination, etc., are all effective in the establishment of the contact surfaces of the bone and implant with maximal intact bone (or minimal defective bone) and also effective in natural performance of the surrounding soft tissue (ligaments, muscle, tendons) to support the implanted site with physiological constrains and loading conditions.

## 8.6 BIOLOGICAL ASPECTS OF BONE-IMPLANT INTEGRATION

Biological integration between the bone and the implant is enhanced by allowing blood flow into the implant surface and adherence of the osteogenic cells to the implant surface. The flow of blood would also increase the density of growth factors at the bone-implant interface to accelerate and facilitate differentiation and proliferation of the osteoblast cells on the surface of the implant. Furthermore, blood flow would increase the density of genes to induce osteoblast cells to create collagen fibrils which establishes the adsorption of the minerals to the osteoblast cells from which osteocyte cells is formed and by gradual maturation of osteocyte cells, the lamellar bone would be structured on the surface of the implant. It could be concluded that by increase of blood flowing on the surface of the implant, its osseointegration is significantly improved via biological activity of the bone tissue at the bone-implant interface. Such biological integration is highly advantageous for long-term implantation of the orthopedic implants to the bone tissue. As highlighted in Chapter 7, porous structuring of the implant surfaces in contact with the bone would significantly enhance the flowing of the blood through the implant surface.

## 8.7 RESTRUCTURING OF THE IMPLANT AT THE BONE-IMPLANT INTERFACE

Through the consideration of bone biological activity and characteristics, implantation of the implant could be promoted by structuring new implants with higher flexibility in mechanical properties and biological interaction with the bone. It might be a dream to expect that a smart mechanical implant could be developed which would mechanically activate the biological performance of the implant. Indeed, by looking into smart mechanical behavior of the bone tissue (according to the Wolff's law) to counteract with transferred loading conditions, developing smart implant that can be biomechanically flexible in interaction with the bone might be achievable. Distribution of stress or extent of load transferring through the bone or implant is not consistent and there would be several localized stress zones. Bone has the ability to restructure to counteract the exceeded effect of loading concentration and even has the ability to reorganize to have higher mechanical modulus in loading direction. This is when such smart restructuring has not been developed in the current biomaterials. Therefore, in bone-implant mechanical interaction, the implant would somehow prevent effective and natural living of the bone tissue by enforcing the bone to be matched with the implant. In fact, the scenario should be reverse meaning that the implant is restructured to better match the bone. The major effect of such dissimilarity between the bone and the implant is observed at the bone-implant contact interface from which the osseointegration is highly affected. As highlighted in Chapter 7, one of the current development concepts in the enhancement of mechanical interaction between the bone and the implant is to establish a super elastic porous layer on the surface of the implant which would cause alteration of surface structure of the implant in contact with the bone and thus, enhanced integration of the bone through the implant. However, such concept might not be applicable for all orthopedic implants to be fabricated from the super elastic porous material. The current smart material which is known as shape memory alloy is to keep the elasticity of the material in high level of strain. This material is the advanced material which would be utilized in the fabrication of various types of implants (heart stent, arthrodesis implant for foot trauma, or osteotomy fixation). However, it is advantageous to develop a material that can be restructured in localized stress to reduce the effect of stress concentration which would necessitate the growth of material technology to invent a high mechanical modulus material with super restructuring ability in interaction with the bone. This might be achieved if the localized stress areas could be reorganized in molecular structuring or molecular bonding. The other concept of implant restructuring might be the changes in the geometry of the implant. Similar to the ability of the bone to become internally denser or weaker or even having external geometry alteration, the implant could have such behavior in excessive localized stresses.

## 8.8 ENGINEERING OF OSSEOINTEGRATION

As reviewed in Chapters 5–7, various types of methods and strategies are being utilized for the enhancement of the osseointegration. The effectiveness of such methods for the promotion of implant bioactivity at the contact surfaces with the bone was investigated and evaluated in relevant published papers. The engineering of osseointegration of the orthopedic implants could be categorized into five groups. In group I, the implant might be coated with an osseoconductive or osseoinductive material to enhance the bone on-growth on the implant contact surfaces with the bone. The second group is associated with porous coating of the osseoconductive or osseoinductive material on the implant surface. In the development of osseointegration in the third group, immobilization of biological factors (e.g., growth factors, proteins) through the porous coated layer was addressed as an effective solution to magnify the osseointegration. In the fourth group, osseointegration has been improved by porous structuring of the implant surface (in macroscale) to allow bone in-growth through the implant surface and thus higher bone-implant integration strength compared to bone on-growth development concept. The other group of methods to enhance osseointegration is the attachment of porous structure to the implant surface through the bonding adhesion method (as reviewed in Chapter 7), which could be considered as the fifth group.

The engineering of these five methods might be managed via various processes in different process conditions. In order to coat a bioactive layer on the implant surface (group I), several methods such as plasma spraying coating, sputter coating, dip coating, electrochemical coating, biomimetic coating, plasma electrolyte oxidation, etc., are utilized. Each coating process would have more than three effective parameters to control and optimize the formation characteristics of the coated layer. Current coating processes are utilized to promote the bioactivity of various types of implants such as femoral stem, tibial tray, pedicle screw, dental implant, etc. Plasma spraying process would be the main method that is industrially utilized for majority of commercial development. Recently, physical vapor deposition methods such as sputter coating has been addressed as the industrial alternative solution to deposit the thin coating layer with nano thickness for those implantation at which the bone and implant are fitted together with force (e.g., intervertebral fusion cage implant which is implanted between two vertebrae). One of the main challenges of such processes is the high cost of the relevant processing machine which would motivate some of the manufacturers to utilize electrochemical coating methods as the low-cost processing method. Due to advantages of porous coating layer for attachment of osteogenic cells compared to coating layer with low porosity, porous coating methods (e.g., plasma electrolyte oxidation) would be more attractive in the enhancement of osseointegration. Porous structuring of the coated layer would be roughly achievable by blasting of the implant surface, however, by alteration of coating parameter under specific pressure and temperature, the coated particles might be structured with the formation of interlayer pores. In order to enhance the biological activity of the osteogenic cells on the surface of the implant, porous structure with effective pore size and pore interconnectivity is processed on the implant surface. Porous layer could be structured directly on the implant surface or structured individually and then attached to the implant surface. The engineering of the porous structure on the surface is carried out using various methods which might be different between the metals and polymers (as reviewed in Chapter 7). Consistent porous characteristics has significant influence on the effectiveness of osseointegration between the bone and the implant, particularly in trabecular bone. Normally, processing methods for the enhancement of osseointegration are initially managed and evaluated on simple shape samples (disc, cylinder) to achieve the optimum processing parameters and conditions. In the next stage, it is attempted to perform the process on the real sample to optimize the conditions and parameters of the final process. Currently, engineering of the osseointegrant solutions is contemplated as the new field of industry in orthopedic implant industry with higher technological processes and equipment.

## 8.9 EVALUATION METHODS OF OSSEOINTEGRATION

One of the main sections of the development process is the evaluation stage. In orthopedic implants, the safety, effectiveness, and performance of the implant is evaluated and validated through various types of biomechanical, biological, and clinical evaluation methods. The evaluation of implant osseointegration could be biologically investigated by in-vitro testing methods to assess formation of bone tissue on the implant surface. However, such biological testing might not have reliable results for the evaluation of bone-implant integration during implantation and postimplantation. The evaluation of implant osseointegration could be associated with the testing of the processed osseointegrant layer. In fact, if the osseointegrant layer is effectively formed, created, or structured, enhanced osseointegration could be expected. Osseointegrant layering would be evaluated in view of adhesion strength and surface characteristics. It is crucial to find the adhesion level of osseointegrant layer with the implant. Adhesion strength should be in an acceptable range to prevent removal of the layer during implantation or replacement of the implant at the intended site. The suggested testing methods for the evaluation of adhesion strength are pull-off and scratch tests. Other than these two methods, the real clinical conditions could be simulated and the implant is implanted into the intended artificial or cadaver bone. Through this test, the osseointegrant layer is tested in real performance and the results are more reliable. Pull-off and scratch tests have some limitations which would reduce the accuracy of judgment on adhesion strength of the osseointegrant layer and implant. One of the main limitation of the pull-off test is using of the adhesive to attach the treated implant to the testing substrate. The adhesive type and its bonding might be different for different processed osseointegrant layer. Likewise, the adhesive may penetrate through the pores of the osseointegrant layer which causes its reinforcement and thus increase the detachment strength which would be higher than the real detachment force. Scratch test may have better reliability in results due to no usage of other material to perform the test. However, in this test, a small scratch is created on the osseointegrant layer which may not represent the adhesion strength of the whole processed layer. Likewise, this test would not be applicable for all types of osseointegrant layering such as surface porous structuring. Due to these limitations or constraints, functional testing protocol might be advantageous for effective and reliable evaluation of adhesion strength of osseointegrant layer. Functional evaluation could simulate the performance of created osseointegrant layer from mechanical aspect of the implantation. Cadaver bone might be a good testing specimen to evaluate the processed implant with osseointegrant layer. The processed implant would be inserted into the cadaver bone according to the intended surgical procedure and then the implant might be removed to perform the surface analysis. During implantation process, a portion of osseointegrant layer might be detached which is measured to achieve the ratio of nondetached section over the whole area of the processed layer. In addition of detachment, the osseointegrant layer is analyzed under high magnification scopes (e.g., scanning electron microscope—SEM) to assess the changes in the structure of the layer such as pore size, interconnectivity of the pores, porous structure profile, level of porosity, wear of osseointegrant layer, etc.

Other than adhesion or structuring strength of the osseointegrant layer, its biological and clinical characteristics are essential to be evaluated. Normally, in-vitro and in-vivo biological tests would show better bioactivity of the processed implant compared to nonprocessed implant. Reliable bioactivity performance of the osseointegrant layering would not be clearly addressed in the literature. In fact, arrangement of such evaluation tests might be very complicated to simulate the real scenario. However, by implantation of the processed implant into the animal body bones or by retrieving of the processed implant from the dead patients or in revision cases, the effective bioactivity of the osseointegrant layer could be better evaluated. As highlighted in Chapter 2, the bone tissue attached to the implant surface and the local blood fluid are taken for biological analysis of the osseointegrant layer. The surface ratio of contact bone to the whole area of processed surface of the implant is measured to obtain the level of bone-implant integrity at the contact interface. The bone tissue attached to the implant is analyzed to find its chemical and physical characteristics such as density, level of organic (collagen), and inorganic composition (minerals). Furthermore, the type of biological factors existed in the extracted blood is determined to assess the extent of bone remodeling at the bone-implant interface. As reviewed in Chapter 2, through the expression level of biological genes (e.g., osteoblastic genes, cytokine genes), proteins, and growth factors, several conclusions could be drawn about the stage of bone remodeling, level of bone formation, proliferation and differentiation of osteoblast cells, formation of lamellar bone, etc. The other evaluation item that could be contemplated is assessment of the bone tissue structuring through high magnification scans (e.g., SEM) to determine the percentage of lamellar bone, osteocyte cells, osteoblast cells, fibrous tissue, defective bone tissue, etc. Through observational investigation of the retrieval implant from the human or animal host, the detached extent of the osseointegrant layer could be evaluated as well. From the exact detachment zones, various conclusions could be drawn about the effective factors that affect the effectiveness and performance of the osseointegrant layer on the whole processed area of the implant. In other words, the osseointegrant layer might have been detached due to different reasons. Detachment or removal of osseointegrant layer might be observed at localized stress zones or at areas that bear higher level of shear stress during implantation or even at areas where there would not be sufficient mechanical load transferring or mechanical strain stimulation. Likewise, the osseointegrant layer could be analyzed on its structural characteristics to clarify any changes or alteration in interval stages of post-implantation. Such assessment could be examined in conjunction with biological investigation outcomes to find the biological-mechanical relations between the extent and the type of formed bone tissue and to examine structure and compositions alteration of the osseointegrant layer. For instance, from biological observation in some areas of bone-implant interface, higher density of mature or lamellar bone could be observed while in those areas there would be no alteration or changes in physical characteristics of the osseointegrant layer.

Detachment of the osseointegrant layer from the implant could be evaluated by examining the adhesion strength in long-term implantation. Although, no mechanical parameter could be quantified for such adhesion strength, the ratio of remained osseointegrant layer over the whole processed layer could be an evaluation factor. From what have been discussed in this section, it could be concluded that evaluation of osseointegration is greatly complicated and various types of histological, mechanical, biological in vitro and in vivo experimental testing, and observational methods are required for completeness and effective examination. Therefore, it would be necessary for the orthopedic implant manufacturers to develop their capabilities in literature science, testing method, and observation techniques for adequate and reliable evaluation of the osseointegrant layering processing.

## 8.10 TECHNICAL GAPS BETWEEN THE SCIENCE AND ENGINEERING OF OSSEOINTEGRATION

The fact that "there would be a gap between the science and engineering" is not new. This is indeed an inevitable fact due to the difference in interests and challenges of each environment. In fact, what is being investigated, studied, or even achieved in scientific environment might not be attractive for the business developers. Engineers may have interest to use the scientists or researchers' achievements in their developing projects, however, the business owners are highly pursuing the market demands rather than introducing a new development method or technology. However, investment in research and development (R&D) projects to explore new technologies and products have been established in top manufacturers of orthopedic implants in developed countries. The R&D department in a company would try to trace and study the universities' discoveries and achievements and use those applicable data, methods, or inventions to the current product range or machinery facilities of the company. Likewise, R&D department attempts to define the new projects based on the market demands or needs and deliver it to the institutes or universities for further investigation or finding out the possible solutions. In terms of discoveries in the enhancement of osseointegration, the current solution is plasma spraying of the hydroxylapatite (HA) or titanium (Ti) on the surface of various types of implants and prostheses which are fabricated from metals (e.g., stainless steel, cobalt chromium, titanium alloys) or polymers (e.g., polyetheretherketone). This is while various types of solution and methods have been investigated (as highlighted in Chapters 5–7) for the promotion of the orthopedic implant osseointegration, however, they have not been commercialized by many manufacturers in the world. One of the key factors to utilize any solution or method to form osseointegrant layer on the orthopedic implant surface is how to create the osseointegrant layer with high level of accuracy and consistency due to the involvement of several process parameters. Furthermore, the evaluation methods of osseointegrant layer would be another matter that produce a large gap between what is achieved in research project and what is expected and required in clinical needs. This might be the other task of R&D department of the manufacturer to attempt to redefine the research projects to enhance the usability and applicability of the research outcomes for real clinical or medical application along with the definition of appropriate testing protocols to assure the safety and effectiveness of the science discoveries.

## 8.11 CLINICAL ASPECTS OF OSSEOINTEGRATION

In terms of clinical aspects of osseointegration, effectiveness of the bone-implant integration in early and long-term implantation could be discussed. Integration of the osseointegrated implants (e.g., percutaneous osseointegrated prostheses for amputees) to the bone is followed-up clinically. In fact, among the orthopedic implants, osseointegrated implants are essentially required to be mechanically and biologically integrated with the bone for long-term functionality. Other types of implants might not be focused post-operatively for their integration with/to the bone as much as osseointegrated implants. This could be found through the published clinical investigations on the topic of osseointegration which report more on the bone-implant integration of the osseointegrated implants rather than on other types of orthopedic implants which are used for fixation, fusion, stabilization, or replacement of various types of bone injuries or disorders. From the clinical view, early and long-term stability of the implant at the implanted position is one of the main clinical demands to achieve the successful clinical outcomes. Integration of the bone and the implant is the key factor for the stability of the implant-bone construct in early and long-term implantation. Therefore, the dislocation, migration, or subsidence of the implant is evaluated in clinical studies. Although, reporting of implant instability is valuable, however, might not be sufficient for clinical evaluation of bone-implant integration or osseointegration of the implant. In this regard, specific and in-depth evaluation methods that can examine the osseointegration (while the implant is in service) are greatly beneficial for the enhancement of clinical outcomes. However, implementation of such evaluation tests might not be effectively feasible. Currently, the effectiveness of the osseointegrant layer is evaluated through the animal studies from which the extent of osseointegration enhancement is compared with one or several reference conditions (normally previous generations of osseointegrant layer are considered for comparison). The results of such studies are beneficial for development, however, would not incorporate the effect of real biomechanical parameters in the enhancement of osseointegration. The other limitation of the animal studies is short-term implantation of the implanted prosthesis and therefore lack of osseointegration evaluation in long-term implantation period. Integration of the implant with cortical and cancellous bones is the other factor that reduces the reliability of the animal studies. In fact, the porosity or compact conditions of the intended use of the implant might not be similar to the animal bone that is implanted. The bioactivity of the osseointegrant layer would be altered in various porosity of the cancellous bone and osseointegrant layer conditions or characteristics would be different in various types of orthopedic implants. Overall, the bone-implant integration is significantly enhanced if clinical conditions are effectively provided postoperatively. In other words, processing of the osseointegrant layer on the implant surface with superior porosity, adhesion, and bioactivity characteristics would not guarantee successful osseointegration of the implant. Various clinical and surgical parameters would influence occurrences and progression of osseointegration process at the bone-implant interface which are reviewed in the following.

### 8.11.1 Postoperative Therapy Plan

Suitable plan of postoperative therapy is beneficial to enhance the bone-implant integration through promoting the physiological interaction between the attached soft tissues (ligaments, muscle, tendon) and bone tissue at the injured site. This would stimulate the required physiological mechanical loading to accelerate the formation of new bone tissue at the intended site. Normally, such therapies are included with the movement of the body organs close to operated site.

### 8.11.2 Bone Quality

Mineral density or level of porosity structuring of the bone, particularly cancellous bone, will have significant effect on the integration of the bone to the implant. For lower density or higher porosity structuring, specific osseointegrant layer is required to facilitate the formation of the bone tissue on the surface of the implant. Likewise, bone remodeling process might not be effectively governed in patient with high level of osteoporosis which necessitates the incorporation of the local drug treatment to induce osteogenesis cells to be formed as mature or lamellar bone at the bone-implant interface.

### 8.11.3 Implant Selection

According to the type and the severity of the injury and patient conditions, various implant types and surgical instruments might be available in the provided system (or set) by the manufacturers to provide higher solutions for the surgeons. Therefore, selection of appropriate implant could enhance the integrity of the implant to the bone at the injured site and thus promote physiological load transferring between the bone and the implant, particularly at the interface areas. To obtain the best solution for the treatment of bone injury, surgeons are required to enhance their knowledge in biomechanical, biological, and material aspects of the orthopedic implants, while engineers and developers need to have the general understanding of the clinical and biological requirements of the orthopedic implants for the treatment of intended bone injury. Such involvement of surgeons and engineers would enhance effective implant selection which significantly improves the clinical outcomes and patient satisfactory level.

### 8.11.4 Implantation Quality

After the selection of the implant, it is crucial to manage the best surgical procedure to obtain the maximal clinical satisfactory level. Surgical procedure for each type of the implant would be different. Likewise, there might be several surgical procedures for one type of implant. On the other hand, for one type of injury and implant, different set of instruments might be developed by manufacturers to enhance the surgical management. Furthermore, the overall surgical procedure would be different such as extensive open reduction, minimal invasive surgery, and limited open reduction. All these varieties in surgical procedures and instrument sets influence the quality of the implantation. Each type of implantation method has its own instrument set. Surgeon would normally consider the time of surgery, patient's general condition, implant type, and instrument set to manage the implantation procedure. Adequate placement of the implant at the intended area would enhance the integrity of the implant to the bone with effective implant-bone preload conditions, implant purchasing to the bone, and bone-implant contact surfaces.

### 8.11.5 Implant Quality

The quality of the implant could influence the bone-implant integration. Today, the number of orthopedic implant manufacturers is increasing and the quality of their implants might be significantly different. The quality of the implant would be affected by the raw material, manufacturing process, precision level of the utilized machines, postproduction processes, quality management system, validation and verification methods, R&D methods, risk management system, design control system, etc. Furthermore, the product features and benefits would have great effects for the enhancement of performance and effectiveness of the implant. Implant must be developed from a raw material that has been rigorously tested in aspects of mechanical and biological requirements of the body. In view of production, implant must be fabricated by high precision machines which are operated by well-trained and experienced technician and engineers. Due to the interaction of the implant surface with body tissues and fluid, surface treatments are necessary to control or enhance the biocompatibility, bioactivity, corrosion resistance, wear resistance, fatigue strength, and nonrelease of toxic elements. Controlling of such interactions is carried out by adequate postproduction processes, which are nowadays considered as one of the main factors of quality in development process.

### 8.11.6 Patient Body Conditions

Body condition of the patient would affect the successful integration of the bone to the implant. It is essential to know the bone mineral density of the patients with osteoporosis. Likewise, the level of minerals in patient's blood is tested to be in reasonable conditions, particularly, in smoker, alcoholic, and elderly patients. The other important patient's body factor is contraction-extension power of the muscles and tensioning of the ligaments and tendons to provide physiological motion conditions at the operated or injured site (such physiological movement would enhance integration of the implant to the bone and promote the required loading stimulation for bone formation at the bone-implant interface). Furthermore, the tolerance of the patient to the operation time may limit the alternative solutions of the implant selection and surgical procedure. Therefore, care must be taken to select the implant with lower surgery period and higher integrity with the bone. In such case, surgeons manage specific and highly controlled postoperative treatment to assure the stability of the implant from which the bone-implant integration might be obtained in longer period than in normal cases.

### 8.11.7 Overall Clinical Issues

Several clinical issues might occur after the implantation of the orthopedic implants. Clinical issues such as infection, soft tissue irritation could affect bone remodeling process at the interface of bone and implant. It is crucial to manage the implantation procedure and to plan adequate postoperative treatments for the prevention of bacterial penetration to the operated site. Likewise, surgical procedure should not cause irritation, damage, or any undesired effect on ligaments, muscles, and tendons at the operated or injured site. Inadequate performance of the local soft tissues would limit the physiological motion at the operated site which would reduce the effective loading stimulation at the bone-implant interface and ultimately affect the osseointegration of the implant.

## 8.12 REMIND AND LEARN

**Q8.1** What development strategies might be contemplated in orthopedic implant industry? What development parameter might be utilized for each strategy?

**Q8.2** Discuss why mechanical integration of the bone and implant might not be sufficient for the effective and long-term osseointegration of the implant.

**Q8.3** Express the benefits of mechanical, biological, and clinical parameters in the development of the orthopedic implants. Provide and discuss the examples of conflicts between biological and mechanical factors in designing and development of orthopedic implants.

**Q8.4** Express the structure of the bone tissue in view of mechanical interlocking with the implant. What is the risk of bone-implant mechanical interlocking? How this risk could be reduced?

**Q8.5** What clinical aspects could be effective in the enhancement of the bone-implant integration? Discuss how each aspect would influence the interaction between the bone and the implant.

**Q8.6** What are the advantages of blood flowing on the implant surface? How blood flowing could be enhanced on the surface of the implant?

**Q8.7** Express the concept of implant restructuring in contact surface with the bone. What are the benefits of such conception? What type of material might be imaginable to achieve this concept? (Explain the characteristics of the imaginable material.)

**Q8.8** What are the categories of the osseointegration engineering? Try to explain each method on the basis of what have been reviewed and discussed in Chapters 5–7. Discuss the effectiveness of each group in the enhancement of the implant bioactivity.

**Q8.9** Express how the osseointegrant layer formed on the surface of the implant could be evaluated biologically and mechanically. What types of mechanical testing methods could be utilized for the evaluation of adhesion strength between the osseointegrant layer and the implant.

**Q8.10** What is the concept of functional testing in the evaluation of osseointegrant layering on the surface of the implant? How the adhesion strength could be evaluated through functional testing? Explain what parameters could be evaluated by performing functional test.

**Q8.11** What types of methods could be utilized for the biological performance of the osseointegrant layer? Discuss the effectiveness of each method in the evaluation of implant osseointegration.

**Q8.12** What are the reasons for osseointegrant layer detachment from the surface of the implant? What types of results could be obtained through the observation of the detached osseointegrant layer from the surface of the implant?

**Q8.13** What is the relation between the alteration in physical characteristics of the osseointegrant layer and effectiveness of the bone remodeling process?

**Q8.14** Discuss the gap between the science and engineering environments in the development of the orthopedic implant and enhanced bioactivity. How R&D department of the manufacturers could increase the usability and applicability of the scientific outcomes in the product development?

**Q8.15** What are the limitations of the animal studies in the evaluation of the implant osseointegration?

**Q8.16** What clinical parameters may influence the effectiveness of the implant osseointegration?

# CHAPTER 9

# Conception of Transition Layer for Enhancement of Osseointegration

## 9.1 INTRODUCTION

From what have been reviewed and discussed in the previous chapters (Chapters 4–7), osseoconductive layer is created on the surface of the implant to enhance the conduction of bone formation on the implant surface (this layer could be called osseointegrant layer). However, this layer would not have a significant effect on load transferring between the bone and the implant. The concept of transition layer could be developed to reduce the mismatch of mechanical properties between the bone and the implant, particularly in metallic implants made of stainless steel, titanium alloy, and cobalt chromium. In metallic implants, the level of generated strain is remarkably low and therefore, the level of required strain for stimulation of the bone formation would not be achieved from which the extent of the bone resorption is exceeded. On the other hand, polymeric implants could generate the high level of strain due to their low elastic modulus (Young modulus). This level of strain would be transferred to the bone tissue during dynamic physiological-loading conditions, which may cause overloading of mechanical stimulation strain for the bone formation and ultimately increasing of the bone deposition or even the bone resorption at the interface of the bone implant. The transition layer concept could be developed to reduce the level of transferred strain to the bone. This layer would be established on the implant surface at the bone-implant surface to be elastically deformed in the adequate level of the strain. Weakening of the bone matrix at the bone-implant interface would cause suppression of the bone through the implant in the long-term postoperative implantation. Due to inertness of the orthopedic implants and bioactiveness of the bone tissue, a bioinert transition layer is advantageous to keep the level of the transferred strain at the bone-implant interface in the long-term implantation. The concept of transition layer is effectively introduced as the

*Osseointegration of Orthopaedic Implants*
https://doi.org/10.1016/B978-0-12-813384-2.00009-6

mechanical transferring layer to improve the mechanical stimulation of the osteogenic cells to be formed as lamellar bone on the implant surface. It is greatly beneficial if the concepts of transition and osseointegrant layer are integrated to achieve the osseointegration pattern with both osseoconduction and mechanical stimulation advantages at the bone-implant interface. In this chapter, such integrated pattern of osseointegration layering on the surface of the implant is hypothesized to open future contributions in this field.

## 9.2 MULTILAYERING OF TRANSITION LAYER

For conception of transition layer, it is worth mentioning that in order to achieve the effective reduction or increase of physiological stimulation strain at the bone-implant interface, the transition layer might be much thicker than the osseointegrant layer (1–3 mm thickness of transition layer vs 10 nm–300 μm thickness of osseointegrant layer). Therefore, the intended orthopedic implant to be featured by the transition layer on its surface is required to be fabricated in lower thickness than the current design. Although such reduction would deduct the mechanical strength of the implant, with consideration of real physiological mechanical and boundary conditions after implantation, optimal performance of the implant would be achievable to develop the implant with the sufficient physiological mechanical strength and the enhanced mechanobiological effect at the bone-implant interface. Transition layer would be an intermediate layer between the bone and the implant with median mechanical properties of the bone and the implant. Furthermore, multilayering of the transition layer would be a major technical need for effective reduction or increase of the strain level. The transition multilayer is structured from two or three materials with the gradual lowering of the mechanical properties. For instance, the transition multilayer on the titanium implant could be organized with modulus of 60 and 20 GPa while the modulus of titanium alloy and cortical bone are 110 and 16 GPa, respectively. This would prevent the concentration of the stress at the interface of the implant core and the transition layer as such might be occurred in the transition monolayer.

## 9.3 TRANSITION-OSSEOINTEGRANT MULTILAYER ON THE IMPLANT SURFACE

Fabrication of the thick transition multilayer or even monolayer on the implant would be a great challenge. Such transition layer might not be processed in porous shape with the target of osseointegration enhancement. It is needed to be in bulk shape structuring as similar as main bulk structuring of the implant. In fact, to achieve the long-term efficacy of the transition layer in transferring the load to the bone, it should not be failed against the fatigue effect. Porous layering would have a reduction effect on the fatigue strength of the implant if the surface of the implant is processed directly for creation of the porous layer. This is while osseointegrant layer is normally a high porous or a coating layer on the surface of the implant and its mechanical deterioration in the long-term implantation would not be contemplated as a design factor. Normally, early or mid-term functioning of the osseointegrant layer is focused to provide early stability of the implant in contact with the bone and long-term function of the osseointegrant layer might not be considered. Due to the high mechanical mismatch between the bone and the implant, the formed lamellar bone might be effected and resorped from which the osseointegration of the implant is weakened. The interesting point that could be highlighted is that the function of osseointegrant layer could be enhanced by mechanobiological transition layer, through which the mechanical mismatch between the implant and the bone is significantly reduced. Therefore, the created lamellar bone on the implant surface could be existed in the long-term implantation. Indeed, this would be a development concept and various types of challenges could be discussed in the development of transition-osseointegrant multilayer on the implant surface. First, attachment or creation of the transition layer might be processed with high-technological processes, which increase the complexity and cost of the processed implant. Second, the bonding strength of the transition layer to the implant would be a major concern for the long-term use of the implantation. In case of a bulky structure of the transition layer, bonding of the transition layer to the implant might be processed by injection overmolding process or be managed by mechanical interlocking. In situation of transition multilayer, adhesion of various layers would be a significant challenge to be processed. Third, bonding of the transition layer must be governed in an isolated area (or cleaning room) to prevent the penetration or trapping of the bacterial microorganisms between the implant and the transition layer. Fourth, good bonding of osseointegrant layer on transition layer would limit the selection of material of transition layer, which needs to be a suitable substrate for coating or processing of osseointegrant layer. Overall, multilayering of the transition layer is highly advantageous for the gradual reduction of stress-shielding effect at the bone-implant interface and for the reduction of risk of osseointegration failure in the long-term implantation period.

## 9.4 COMBINED BULK-POROUS TRANSITION LAYER

Combination of bulk and porous structures in the transition layer might be advantageous for either the gradual reduction of the stress shielding effect and the creation of a transition-osseointegrant multilayer on the surface of the implant. In fact, the bulk portion of this double-structured layer is to bond with the implant surface and porous portion is to bond with the osseointegrant layer (osseointegrant layer might be a coating of bioactive material). This hypothesized transition layer might be processed from one material that would provide the integrity of the bulk and porous sublayers compared to processing of the bulk and porous sublayers from different materials from which their integrity and bonding interface might not be achieved effectively. For processing of the bulk-porous transition structure, first the bulk layer with adequate thickness (about 1–3 mm) is fabricated and then a portion of the bulk layer thickness is processed in porous structure. After creation of bulk-porous structure, the transition layer is achieved by an attachment of the structured bulk-porous layer to the surface of the implant. Attachment process might be achieved by heating of the lower surface of the bulk portion close to the melting temperature followed by penetrating the heated portion to the precreated notches on the implant surface.

One of the advantages of the bulk-porous transition layer is increasing the hydrophilic characteristics of the implant surface. This would enhance the adhesion of the osteogenic cells, growth factors, and proteins to accelerate the formation of the bone tissue on the surface of the implant. Likewise, coating, deposition, or immobilization of bioactive materials (e.g., hydroxyapatite, tri calcium phosphate, chitosan, bioglass, etc.) or biological factors (e.g., growth factor, proteins, etc.) through the porous portion of the transition layer would be effectively achievable compared to nonporous or bulk structuring of the implant surface. In addition, the bulk-porous layer could be considered as the multitransition layer at which the modulus of the transition layer is reduced from the bulk portion through the porous portion. This modulus reduction would deduct the mechanical mismatch between the bone and the implant.

Bulk-porous transition layer could have some disadvantages. First, the porous portion may introduce concentration of the stress at the porous pores from which the crack might be initiated from highly concentrated zones and propagated through the bulk portion. This would reduce the fatigue strength of the bulk-porous transition layer and due to low thickness of the bulk portion, the transition layer might be fractured from which the bone-implant integration is partially or entirely collapsed. This effect of disadvantage might be reduced by increase of the bulk thickness or by using of the superior elastic material (e.g., nitinol) in the fabrication of such transition layer.

## 9.5 MECHANICAL EFFECTIVENESS OF THE TRANSITION LAYER

Transition layer could be developed as a long-term osseoconductive layer with a mechanical stability on the surface of the implant. In such a type of transition layer, the prominent factor is bioactivity rather than mechanical interface layer for reduction of mechanical modulus mismatch between the bone and the implant. One of the main challenges of osseointegrant layers that has been reviewed and discussed in the previous chapters is the deterioration of mechanical stability and bioactivity. Therefore, it is essential to develop a new concept to enhance the mechanical effectiveness of osseointegration at the interface of the bone implant. Overall, osseointegrant layering is developed by coating, deposition, or creation of bioactive particles or factors on the implant surface. On the other hand, porous structuring of the implant surface would have a significant effect on the adhesion of biological factors to the implant surface from which the bone formation is conducting on the implant surface. During the period of postimplantation healing, the partial (not full) physiological-loading conditions might be transferred to the bone-implant interface, which allows biological formation of the new bone tissue at the interface. By ending of the healing period, patient would be advised to have a normal life and do the daily activities with some limitations. This would introduce full natural physiological loading to the interface which may affect the stability of created bone tissues on the implant surface and weakening of osseoconductivity of the implant surface. By this introduction, the matter is how to establish an osseointegrant layer with the ability of restructuring, reorganizing, or recharging of biological factors at the bone-implant interface. In case of the porous osseointegrant layer, through cyclic or dynamic effect of physiological-loading conditions which is generated in daily activities, the pores of porous structure might be deformed or even destroyed from which osseointegration of the implant is affected. Therefore, restructuring of the porous structure would not be feasible at all. In case of recharging of biological factors, there might be a possibility to inject biological factors on the implant surface to induce conduction of bone reformation at the affected area in the long-term postimplantation. However, through the injection, the biological factors (hydroxyapatite, chitosan, protein, stem cells, osteoblastic genes, etc.) would be spread or located at the bone-implant interface and not to be deposited with the high adhesion strength as can be achieved at the processing phase of the implant fabrication. Since the bone and the implant are normally in compressed conditions, injection of the biological factors at the interface would be deposited by immobilization of the treated site to allow activity of the biological factor for a period of time.

## 9.6 MAGNESIUM-POLYETHERETHERKETONE (MG-PEEK) TRANSITION

A combination of PEEK and magnesium was first addressed by the author in the book "Trauma Plating System." The concept was hypothesized for the development of the trauma plate and screws for fixation of bone fractures. Due to close mechanical properties of the PEEK to the cortical and trabecular bones, using of this advanced polymer in the transition layer could have several advantages for the enhancement of load transferring between the metallic implants and the bone. Likewise, magnesium and magnesium alloys are biodegradable and attractive to be used as the osseoconductive layer on the surface of the implant. As reviewed in Chapter 7, surface of the PEEK could be processed in porous form. This availability of processing technology would attract processing of PEEK as the transition layer with two sides of porous layering. As illustrated in Fig. 9.1, one porous side of PEEK transition layer is integrated to the porous surface of the implant, while the other

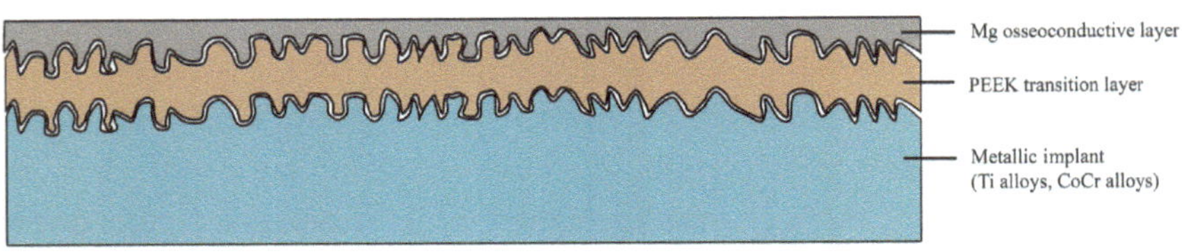


**Fig. 9.1** Schematic illustration of the magnesium-polyetheretherketone (Mg-PEEK) transition layer for bioactivity enhancement of the metallic implants.

porous side of the PEEK transition layer could be integrated to the magnesium (Mg) osseoconductive layer. By degradation of the Mg, bone formation is accelerated with complete degradation of the Mg layer, the bone matrix is penetrated to the pores of the porous PEEK from which the osseointegration of the transition layer is accomplished. With adequate integration of the porous PEEK transition layer to the implant, the long-term osseointegration with advantage of mechanical load transferring between the bone and the implant is obtained.

## 9.7 SUMMARY

The two general methods of bioactivity enhancement as coating of osseoconductive or osseoinductive materials and porous layering on the surface of the orthopedic implants have been extensively investigating in both scientific and industrial development sectors to form an osseointegrant layer on the surface of the orthopedic implants. As the third conception in enhancement of the implant osseointegration, transition layer was introduced in this chapter. Transition layer is created in higher thickness than the osseointegrant layer to establish an intermediate layer for enhancement of load transferring between the bone and the implant. Transition layer could be structured in single or multilayers. The transition multilayer could be created in bulk or bulk-porous form. Likewise, the pattern of transition-osseointegrant layer could be developed to form the transition layer with coating or deposition of an osseoconductive or osseoinductive material on its surface. Furthermore, the concept of porous-bulk-porous processing of the PEEK transition layer with deposited magnesium alloy has been hypothesized as the novel idea for the long-term enhancement of the osseointegration of the metallic orthopedic implants. Overall, the concept of transition layer might be a breakthrough concept for the development of orthopedic implant with greater osseointegration characteristics in early and long-term implantation.

## 9.8 REMIND AND LEARN

**Q9.1** What are the main categories of bioactivity enhancement of the orthopedic implants?

**Q9.2** Express the concept of transition layer for the enhancement of the implant osseointegration in the long-term implantation?

**Q9.3** What are the main characteristics of the transition layer?

**Q9.4** Discuss how multilayering of the transition layer would be beneficial in enhancement of the osseointegration.

**Q9.5** What is the concept of bulk-porous transition layer? What are the advantages and disadvantages of such conception compared to entire bulk transition layer?

**Q9.6** What is the concept of transition-osseointegrant layer?

**Q9.7** Explain the concept of Mg-PEEK transition layer. Discuss how this concept would be processed and how could it be beneficial for the enhancement of the osseointegration in the long-term implantation.

# INDEX

Note: Page numbers followed by *f* indicate figures and *t* indicate tables.

## O



## P

FSC
www.fsc.org
MIX
Papier | Fördert
gute Waldnutzung
FSC® C083411

Kolibri 360 GmbH
Ferdinand-Jühlke-Straße 7
99095 Erfurt, Deutschland
produktsicherheit@kolibri360.de